Coconut and Oilpalm

Genetic Upgradation and Protection

T. Ramanathan

2018

Daya Publishing House®

A Division of

Astral International Pvt. Ltd.

New Delhi - 110 002

First Published, 2001
Reprinted, 2018

ISBN: 978-81-7035-257-0 (Hardbound)

Publisher's note:

Every possible effort has been made to ensure that the information contained in this book is accurate at the time of going to press, and the publisher and author cannot accept responsibility for any errors or omissions, however caused. No responsibility for loss or damage occasioned to any person acting, or refraining from action, as a result of the material in this publication can be accepted by the editor, the publisher or the author. The Publisher is not associated with any product or vendor mentioned in the book. The contents of this work are intended to further general scientific research, understanding and discussion only. Readers should consult with a specialist where appropriate.

Every effort has been made to trace the owners of copyright material used in this book, if any. The author and the publisher will be grateful for any omission brought to their notice for acknowledgement in the future editions of the book.

Published by : **Daya Publishing House®**
A Division of
Astral International Pvt. Ltd.
– ISO 9001:2008 Certified Company –
4736/23, Ansari Road, Darya Ganj
New Delhi-110 002
Ph. 011-43549197, 23278134
E-mail: info@astralint.com
Website: www.astralint.com

ACKNOWLEDGEMENT

The author wishes to place on record his sincere thanks to Mr. P.K. Thampan, who championed the cause for coconut development for (1) readily accepting to write a foreword for this book and (2) generously granting permission to use data and information contained in the book "Coconut for Prosperity" edited by him. The author's grateful thanks are also due to Dr. M. Subramanian, then Director, Tamil Nadu Rice Research Institute, Aduthurai for providing facilities to take photographs of oil palm. He is indebted to M/s Daya Publishing House, Delhi for meticulous care in printing and prompt release of the book with nice get up.

T. Ramanathan

FOREWORD

Coconut and Oilpalm are the two major tree crops grown extensively in the wet tropical regions of the world. Coconut cultivation is mostly in the hands of small farmers who depend on the multiple products derived from the versatile palm for their livelihood security. Oilpalm, on the other hand, is a single product crop and is favoured for cultivation in larger holdings and plantations. While the edaphic and climatic requirements of both the crops are more or less similar for optimum growth and productivity, coconut exhibits better tolerance to less favourable conditions, a trait appreciated by small farmers. Likewise, coconut processing is manageable even at the farm-household levels whereas special facilities are to be created for processing oilpalm fruits. May be for these reasons oilpalm has gained only a marginal spread in India despite its first introduction in the fagend of 19th century.

Research on coconut in India had its beginning in 1916. Since then, commendable achievements have been made in the spheres of crop improvement, production and protection. The first introduction of exotic cultivars from different countries was made in 1924 for comparative evaluation with indigenous types. Likewise, the phenomenon of heterosis or hybrid vigour in coconut was first noticed in 1932 and in as early as 1936 the production of high yielding hybrids was commenced. This was followed by studies on the genetics of dwarf, prepotency, genetic correlation of phenotypic characteristics of palm with yield, hybridization technique, nursery management, mineral nutrition, cultural management, water management techniques, farming systems and pest and disease management. The valuable findings emanated from these studies are being made use of in many coconut growing countries for the development of their indigenous coconut industry.

Notwithstanding the achievements already made, efforts are now needed to develop high yielding and drought tolerant coconut varieties for cultivation in problem areas. Likewise, better hybrid combinations which are devoid of undesirable traits such as alternate bearing, bunch buckling, susceptibility to marginal water stress etc. are to be evolved and made available to the farming community. Techniques for the suppression of major pests and diseases biologically are also to be developed.

In the above context, a review and assessment of the research on coconut improvement and protection become very useful. A similar approach to cover oilpalm will also prove to be rewarding. The book 'Coconut and Oilpalm : Genetic Upgradation and Protection' is expected to serve the desired purpose. The author of the book Dr. T. Ramanathan has to his credit many years of experience in the relevant areas. This has helped him to present an overview of the ongoing research on coconut and oilpalm in the major producing countries with special reference to the achievements made in India. I congratulate Dr. T. Ramanathan for bringing out this useful book which will be much appreciated by students and researchers both within and outside the country.

Gandhi Nagar,
Cochin

P.K. Thampan
Peekay Tree Crops Development Foundation

Preface

Oil seeds form the second largest agricultural commodity after cereals in India sharing 13 per cent of the country's gross cropped area. Despite the increase in oil seeds output from a level of 12-13 million tonnes till 1987-88 to 17 million tonnes at present, the demand for edible oil still falls short of the indigenous supply to the extent of a million tonnes and it has to be met, through imports, thus causing a drain on the scarce foreign exchange resource of the country. Cultivable land is shrinking gradually due to conversion of land as house sites for accommodating ever growing population and also increased encroachments for industrial uses. Thus, the need for increasing productivity of crop plants, particularly oil seeds through genetic improvements is felt to a greater extent now than ever before.

Coconut and oilpalm are the two sources of edible oil from perennial trees which could be relied upon in bridging the gap between supply and demand.

A two pronged approach is essential – one is to enhance the productivity through genetic manipulation and the other to make it sustainable by adopting proper management practices to avoid yield loss due to occurrence of pests and diseases. Hence, concerted attention is paid in presenting information on genetics and breeding in chapter 1 for coconut and chapter 3 for oilpalm. The protection technologies dealing with the management of pests and diseases are covered in chapter 2 and 4 respectively for these two crops.

Coconut : Progress of crop improvement work in various countries like Indonesia, Philippines, Sri Lanka, Jamaica, Ivory Coast (Africa), and India was reviewed and results of research presented. Exploitation of heterosis for greater expression of productivity in terms of nut, copra and oil has made significant progress in recent years, despite the serious constraint of lack of homozygous parent material involved in the hybridization programme.

Progress of research at CPCRI (Central Plantation Crops Research Institute), Kerala, RARS (Regional Agricultural Research Station) Pilicode (Kerala), CRS (Coconut Research Station), Veppankulam (Tamil Nadu), and various Co-ordinating Centres located at Ambajipet in Andhra Pradesh and Ratnagiri in Karnataka has been presented. The role and invovlement of indigenous and exotic cultivars in both tall and dwarf types for hybrid production and utilisation for increased productivity has been narrated. Nearly 75 per cent of coconut area depend entirely on monsoon rains. Physiological basis for drought tolerance and production trends as presented in this book may gain relevance in this context.

Identification of prepotent coconut trees assumes greater importance in the context of mass multiplication of super elite palms and also in the choice of ideal parent combinations for evolution of successful hybrids. Refinement of currently available techniques may lead to greater success in clonal propagation of the usually seed propagated coconut to realise the dream project of achieving chip of the old block of super elite palms.

Somatic embryogenesis still remains untapped to full extent in coconut for several technical reasons, despite of more than two decades of research. Exploitation of embryo culture technique will greatly help in building up large number of germ plasm collection of exotic and indigenous origin as this will reduce the problem posed by transportation of seed coconut in large numbers over long distance.

Pests and diseases affecting coconut are presented in greater detail in Chapter 2. The pests include insects, mites, nematodes and rodents. Redpalm weevil, rhinoceros beetle, black headed caterpillars are the major insects that cause considerable damage, of which the first one often proves to be fatal. Coconut scale, white grub, bagworm, termites, mealy bugs, coreid bug, scolytid bark borer and weevil Diocalendra also have a role to play in destabilising the trees. Ecology, nature and extent of damage, biology of the causal agent, control measures which include, biological, chemical, cultural and integrated pest management (IPM) are furnished, especially for major pests which are of serious concern to the farmers.

The second part of chapter 2 deals with diseases of coconut. Among the several diseases affecting coconut, Thanjavur wilt is most serious, devastating and fatal. Root (wilt) is a debilitating disease often leading to uneconomical returns. Stem bleeding, bud rot, leaf blight, button shedding and nutritional disorders including tapering wilt are other diseases that assume serious proportion occasionally. Details are given to the extent needed, on the causal agent, pathogenicity, symptomatology, spread of the disease, and control measures. Use of chemicals, antagonistic micro-organisms, vector control in the case of root (wilt) caused by MLOs, and integrated management practices are outlined. Trees affected by nutritional disorders are recovered after appropirate inputs of macro and micro nutrients.

Oilpalm is credited with higher oil productivity (4 to 6 tonnes/ha) compared to less than two tonnes in coconut. Malaysia and Indonesia have a major share in the world's palmoil production. Though the crop was introduced in India in the fag end of 19th Century (in Calcutta) it took more than 60 to 70 years in 20th Century to see the same on plantation scale in Kerala. Plantations were established in 80s and 90s in Andhra Pradesh, Karnataka, Maharashtra and to a lesser extent in Tamil Nadu.

Unlike coconut, which gives fairly good yield even in drought situation, oilpalm does not tolerate drought condition. Under prolonged drought condition female inflorescences seldom emerge and to the contrary, male inflorescences are produced in abundance under such situation. The converse, i.e., production of mostly female inflorescence is also true in the case of Indonesia and Malaysia where the annual rainy days exceed 200.

Oilpalm research is still in the infant stage in India. Intensification of research efforts are needed in this important edible oil yielding crop. In chapter 3, care was taken to provide basic information on the plant product and uses of palmoil, besides furnishing details on taxonomy, morphology, methods of raising nursery and planting. Genetic resources are meagre in this crop.

Breeding objectives and methodologies are narrated. Progress made in marker assisted selection, RFLP technique, identification of transgenic palms and clonal multiplication techniques are outlined.

Achievements made in somatic embryogenesis and embryo culture are included. Elsewhere, efforts were also made in breeding for short stature, and ideotype. Occurrence of stem-rot in the Far East, Vascular wilt in W. Africa and sudden wither in S. America led to breeding for resistance to these diseases abroad.

Insect pests, birds and rodents affecting oilpalm are listed in Chapter 4, Diseases affecting nursery and adult palm are also furnished in this chapter. Results of research on processing technology undertaken at Kumulur (Tamil Nadu) included in this chapter is of particular importance in the quality aspect of palmoil.

ARNI-632 301

T. Ramanathan
Former Professor and Head
Department of Oil Seeds
Tamil Nadu Agricultural University
Coimbatore - 641003

CONTENTS

Introduction

Vegetable Oil Scenario

Global

World Production of oils and fats was about 85 million tonnes in 1994 of which 64 million tonnes is vegetable oil. The animal fats (butter, lard, tallow and fish oils) proportion of world production has fallen from one third to one quarter over the past decade.

The main trend in vegetable oils consumption has been growth in consumption due to switch from dairy and other animal fats and also to higher levels of disposable income in developing markets. Some vegetable oils have benefitted more than others in the USA sales of less saturated oils such as rapeseed oil have increased market share at the expense of tropical oils such as palm oil which are high in saturated fats. Similarly, in Europe, consumption of sunflower oil has benefitted from health concerns. Main sources of vegetable oils are Soyabean, Palm and Palm Kernel, Sunflower, rapeseed, cottonseed, groundnut and coconut oil. The dominant edible oils–soyabean, palm, rapeseed and sunflower seed–account for over 70% of world production. During 1980s Palm, Palm Kernel, Coconut and rapeseed have been gaining share while soyabean has been contracting. At the same time there has been a shift in regional production patterns in Brazil, Argentina and China increasing share of soyabean production, India and Poland increasing share of rapeseed production and EC increasing share of both rapeseed and sunflower seed production. During the 1980s global output of vegetable oils increased at a rate of just over 4% per annum and there have been substantial changes in the proportion of output accounted for by the different types of vegetable oil (Oil & Fats International file 1995-2000).

National

Although India has 20.8 per cent of the world's area under oilseeds, it accounts for less than 10 per cent of world's production to meet the needs of about 16 per cent of the world's population. Oil seeds form the second largest agricultural commodity after cereals in India sharing 13 per cent of the country's gross cropped area and accounting for nearly 5 per cent of gross national product and 10 per cent of the value of all agricultural products.

No other country has its range of perennial and annual oilseeds. In terms of area, India ranks first in groundnut, sesame, linseed, safflower, niger and castor, second in rapeseed-mustard and third in sunflower.

In general, the production of all oil seeds showed a rising trend from 1985-86 to 1994-95. The increase was sizable in groundnut where production rose from 5.12 million tonnes in 1985-86 to 8.40 million tonnes in 1994-95, an increase of 64 per cent. Other oil seeds where sharp improvements were seen during this period are rapeseed-mustard (130 per cent), sesame (68 per cent), sunflower (404 per cent), soyabean (260 per cent) and castor (181 per cent). The crops with modest increases in production are safflower (34 per cent), linseed (16 per cent) and niger (5 per cent).

The achievements of the targetted production of 22.48 million tonnes of oil seeds by 1994-95 as against 10.83 million tonnes in 1985-86 is an indication of sustainable efforts by all the agencies concerned (Reddy, 1995).

Perennial Oil Bearing Trees

India has a variety of perennial oil yielding trees which are capable of yielding higher tonnage of oil per hectare. Of these only a few have been exploited fully for their oil, while many are under-exploited or unexploited. About 30 tree species like palms, plantation crops, avenue and forest trees and minor forest trees grown in India yield edible and industrial oils. These can definitely yield substantial quantity of edible/industrial oil if properly exploited.

Of the above, coconut (*Cocos nucifera* Linn) with wide collection of 86 exotic and 41 indigenous germ plasm which helped to develop eleven varieties and hybrids, has been contributing about 0.3 million tonne of oil annually. This crop being cultivated in 1.5 million hectares producing 9700 million nuts, and additional potentialities exist for 0.496 million hectares in both traditional and non-traditional areas. The germ plasm evaluation and hybridization have resulted in the release of two varieties and nine hybrids which can yield 1.5 to 3.0 tonnes of oil per annum.

A recently introduced crop, oilpalm (*Elaeis guineensis* Jacq) is another source of vegetable oil which can yield three to six tonnes of oil per hectare per annum has a limited germplasm of nine exotic collections. While available *dura* and *pisifera* could be used to produce about three lakh tonnes of hybrid seeds, the bulk of seed supply is from outside the country. There is a potential for growing about 0.7 million hectares under irrigation in eleven states, and during Eighth Five Year Plan alone it was envisaged to increase the area under production to about 88,000 hectares.

A list of oil bearing trees are furnished in Table 1.1.

Table 1.1 : List of oil bearing trees and their Oil content (Rethinam, 1993)

S.No.	Common Name	Botanical Name	Oil Source	Oil Content (%)
1.	Coconut	*Cocos nucifera* Linn	Kernel	63-70
2.	Oilpalm	*Elaeis guineensis* Jaq	Mesocarp	21
3.	Macauba	*Acrocomia sclerocarpa*	Mart	–
4.	Tucuma	*Astrocaryum tucuma* Mart	Pulp Kernal	35-38 45-52
5.	Pinhao/Physic nut	*Jatropha Curcas* Linn	Seed	40
6.	Piqui	*Caryocar brasilience*	Pulp Kernel	67 70
7.	Jojoba	*Simmodsixa Chinensis* Link	Schneider	45-50
8.	Sal	*Shorea robusta* Gaertn F	Fat	19-20
9.	Mahua	*Madhuca indica* J.F. Gmel	Kernel	50
10.	Neem	*Azadirachta indica* Ajuss	–	45-50
11.	Pungam or Karanj	*Pongamia pinnata* Pierre	–	27.30
12.	Cashew	*Anacardium accidentale* Linn	Kernel Shell	35-40 25-30
13.	Mango	*Mangifera indica* Linn	Kernel	10
14.	Almond	*Psumus amyodalus* Linn	Kernel	38-55

Macauba (*Acrocomia sclero-carpa* Mart), Tucuma (*Astrocaryum tucuma* Mart), Pique (*Carycoar* Linn), Pinhao (*Jatropha Curcas* L.) etc. are promising oil yielding trees capable of yielding 1.5 to 7.0 tonnes of oil which need to be introduced and multiplied (Rethinam, 1993).

1

Coconut (*Cocos nucifera* L.)

Among the perennial oilseed crops, coconut is most ubiquitous and demands a serious attention by agriculturist and food technologists. Coconut palm which has the highest economic value, is grown over the widest geographical area and has possibly the largest number of uses and is most suitable for use in wide variety of foods. The Government of India has recognised coconut as a tree-based oilseed crop and has given it the status of oilseeds for the purpose of trade and commerce. Coconut ranks seventh in position in world fats and oil production.

India has emerged the largest producer of coconuts in the world, though only third in area (after Indonesia and Phillipines) under the crop. According to 1995 data, production was 13,300 million nuts accounting for 24.6 per cent of the global output from an area of 1.7 million ha which represented 15.2 per cent of the global area devoted to coconut.

Table 1.2 : Area and Production of Coconut in the World

Country	Area (000 ha)	Share %	Production (million nuts)	Share %
Indonesia	3,712	32.83	12,961	24.02
Phllipines	3,164	27.99	11,700	21.68
India	1,714	15.16	13,300	24.65
Sri Lanka	419	3.71	2,755	5.10
Thailand	412	3.64	1,898	3.52
Malaysia	290	2.57	748	1.39
Papua New Guinea	260	2.30	869	1.61
Vietnam	186	1.64	1,000	1.85
Other Countries	1,149	10.16	8.732	16.18
Total	11,306	100.00	53,963	100.00

Source : APCC Statistical Year Book 1995 and updated figure for India by Coconut Development Board (Thampan, 1997).

Among the major coconut growing states, the share of Kerala in production was 42 per cent followed by Tamil Nadu with 31 per cent, Karnataka with 10 per cent and Andhra Pradesh with 8.8 per cent.

Table 1.3 : Area, Production and Productivity of Coconut in India for 1995-96

State/Union Territory	Area ('000 ha)	Production (million nuts)	Productivity (nuts/ha)
Kerala	982.1	5,905.7	6,013
Tamil Nadu	298.6	4,345.7	14,554
Karnataka	269.4	1,406.5	5,221
Andhra Pradesh	90.0	1,231.4	13,682
Goa	24.7	119.0	4,818
Orissa	42.9	234.5	5,466
West Bengal	23.1	279.4	12,095
Andaman Nicobar islands	24.4	85.4	3,500
Assam	17.8	126.2	7,090
Maharashtra	8.2	169.1	20,622
Tripura	9.4	4.7	500
Lakshadweep	2.8	26.5	9,464
Pondicherry	2.1	33.8	16,095
All India	1795.5	13,967.9	7,779

Source : Thampan, 1997.

GENETICS AND BREEDING

Researches on varietal improvement in plantation crops are beset with a variety of problems such as narrow genetic base, long Juvenile phase, exclusive seed propagated nature and highly heterogenous nature of the phenotypes involved. Coconut is not an exception to this general observation.

In coconut, selection procedures adopted have yielded varieties like Lakshadweep ordinary. Coconut hybrids involving different dwarfs and Talls have yielded about 130% increase compared to the best parent. In addition, naturally occurring super palms have been identified. A quantum jump in the yield of coconut from its present national average of 30 nuts to even 200 can be achieved if an efficient method of tissue culturing of such palms can be standardised. The monotypic nature of coconut will, however, continue to be a limiting factor in breeding coconut varieties suitable for cultivation in diverse ecological situations (Bavappa and Nair, 1989).

The centre of diversity of coconut is in the South-East Asia and Malaysia. According to FAO statistics, 58.1% of the six million hectares under this crop in the world is in South-East Asia. Another 24.6% of the area occurs in South Asian countries.

Dwyer (1938) described some methods for coconut improvement. They consist of :

(1) Introduction of varieties.

(2) Artificial self-pollination.

(3) Variability and germplasm collection.

(4) Hybridisation – natural or artificial.

(5) Mass selection and strain building.

(6) Close and line breeding.

(7) Plant-to-row method.

1. Introduction of Varieties

Introduction of promising varieties in different parts of the world is a profitable method of crop improvement. German government introduced superior kinds of coconut, notably Nice Vai and Nice ice from Samora into the Rabul Botanic Gardens, New Guinea as early as 1911. The introduction of even outstanding varieties or forms from distant countries for large scale plantation has to be done with circumspection, because all varieties are not cosmopolitan and may not perform equally well in different ecological conditions. The famous 'San Blas' coconut of Panama was disappointing when introduced into Malaya. Introductions should be restricted to varieties which are distinct and superior so that they are not duplications of existing ones.

2. Self Pollination

According to Harland (1957) since coconut palm is a cross fertilized plant, it has almost certainly accumulated by mutation unopposed by selection, a large number of deleterious recessives, and these would come to light on selfing. He has indicated that deleterious recessives would also occur in paired crosses, since the parents might be heterozygous for the same recessive gene. Jack and Sands (1929) were of the opinion that the dwarf variety of coconut showed greater definiteness and generally bred true to type because of the fact that it was commonly self-fertilized.

In Dwarf variety, self pollination is the general rule due to the overlapping of the female and male phases in the same inflorescence. Self pollination occurs in certain seasons in the Tall variety also. For artificial selfing, pollen is collected from the same inflorescence or from the next one according to availability.

3. Variability and Germplasm Collection

The Talls in general, are cross pollinated and exhibit a wide range of variability for height of the palm, shape of crown, colour, shape and size of nuts, yield and annual out-turn of copra. This could possibly be one of the reasons for identifying coconut varieties, especially the Talls, by the location of collection/ cultivation. In addition, some of the cultivars are also known by the special features they possess, such as Lakshadweep Micro (a small-fruited variant of Lakshadweep Tall), Niu Bulundrau (producing 100 coconuts per bunch, from Fiji) and Kappadam (from Kerala having large quantity of nut water). The dwarfs on the other hand, are distinguished by the colour of the nut, and location of their general occurrence (e.g. Chowghat Dwarf Orange, Malayan Dwarf Yellow/Orange/Green and Cameroon Red Dwarf).

The first systematic classification of coconut varieties and forms was attempted by Narayana and John (1949). Subsequently, Gangolly *et.al.* (1957) reviewed the literature on coconut varieties. Liyanage (1958) classified Ceylon (Sri Lanka) coconut varieties. The criteria employed for all these classifications were, the plant habit, Juvenile period and geographical source. The third classification was proposed by Fremond *et.al.* (1966) on the basis of floral biology. Harries (1978) while reviewing the evolution, dissemination and classification of coconut has proposed the use of fruit component analysis data as a rational approach to the classification of coconut varieties. Germplasm evaluation undertaken in different countries is summarised below.

Indonesia

Until, 1991, in Mapanget Experimental Garden, north Sulawesi, there were 35 coconut cultivars consisting of 10 cultivars of Dwarf coconut and 25 cultivars of Tall coconut which were collected from North Sulawesi, Central Sulawesi, North Sumatra, Bali, Muluka, West Java and East Java. In Pakuwon Experimental Garden, West Java, there were 22 coconut cultivars consisting of cultivars of dwarf coconut and 14 cultivars of tall coconut. In Bone-Bone Experimental Garden, South Sulawesi, there were 41 coconut cultivars collected from various regions in Indonesia. However, due to the new organisational structure, both Pakuwon Experimental Garden and Bone-Bone Experimental Garden are no longer attached to the Coconut Research Institute. Some of the coconut cultivars were used as the parent plants in the production of improved types (Michellia Darwis, 1996).

Philippines

Botanically speaking, there are five varieties of coconut : Typica, Androgena, Spicata, Javanica and Nana. The most common forms or types of tall palms growing in the Philippines are San Roman, Laguna, Makapuno, Lono, Taban or Cuyamis, Tataguden, Bontanun, Lupisan, Tutupaen, Carahunor Saloquihan, Agta, Busag, Bulao and Palawan. The dwarf varieties found growing in the Philippines are Coco nino, Pugai, Dahili and Inano and Mangipod (Iznaga and Tabayoyong (Eds), 1967).

Sri Lanka

Manthriratne (1972) presented data on growth, flowering and yield of dwarf green, dwarf yellow and dwarf red coconut palms. The shortleaf span of this variety makes it possible to grow about 227 palms per hectare. All three colour forms bear early when compared with the commercially grown 'typica'. However, the dwarfs are a failure as a plantation crop for they yield not more than 1.25 - 2.50 tons of copra per hectare because nuts are small and overall cost per unit copra is high. The dwarfs yield 85-113 g copra per nut. The dwarf variety constitutes a valuable source of breeding material on account of its early bearing character.

Jamaica

The introduction of the Malayan Dwarf coconut, its performance in Jamaica are briefly reviewed by Harries, (1970). Due to its precocity, this variety is less susceptible to hurricane damage than the Jamaica Tall variety. All three dwarf types (red, yellow and green) are resistant to lethal yellowing disease. The variety starts bearing after 3½ years and its ripening period is 11 months. Although 5,000 to 9,000 dwarf nuts make one ton of copra compared to 4,000 to 6,000 tall nuts, and the oil content is slightly lower, the total yield is higher due to higher production per palm and closer planting densities. However, Malayan Dwarf seems more susceptible to leaf spots than Jamaica Tall.

The local tall variety of coconut in Jamaica appears to be identical with tall varieties in other Caribbean islands and on the Caribbean Coast of Central and South America and with varieties in Africa, India and Ceylon. Another tall variety which was introduced to Jamaica occurs on the Pacific Coast of Central and South-East Asia. The presence of "Malayan Dwarf" in the region is a valuable safeguard against the spread of lethal yellowing. (Harries, 1971).

International Coconut Genetic Resources Network (COGENT)

COGENT is a global research network organised for and on behalf of coconut producing countries. It now has 30 coconut producing countries as members. It is supported by the member countries, partner research institutions, donor agencies and international development organisations.

The goal of COGENT is to enhance coconut production on a sustainable basis in order to improve the coconut based economy of the developing countries through improved cultivation of coconut palm and efficient utilisation of its products. The purpose of COGENT is to develop and implement an international mechanism to co-ordinate research activities of national, regional and global significance, particularly in germplasm exploration, collection, conservation and enhancement. It also aims to establish a basis for colloboration on the broader aspects of coconut research and development. (Pons Batugal, 1996). A total of 665 accessions of coconut germplasm is being maintained as per CGRD (Coconut Genetic Resources Data base) in different countries.

Table 1.4 : COGENT Member Countries

South-east and East Asia	South Asia	Pacific	Africa	Americas
China	Bangladesh	Cook Islands	Cote	Brazil
Indonesia	India	Fiji	d'Ivoine	Costa Rica
Malaysia	Pakistan	Kiribati	Ghana	Cuba
Myanmar	Sri Lanka	Papua New Guinea	Kenya	Jamaica
Philippines		Solomon Islands	Nigeria	Trinidad
Thailand		Tonga	Sychelles	
Vietnam		Vanuatu	Tanzania	
		Western Samoa		

India

The cultivated varieties and their relatives which are the source material for crop improvement programmes are being maintained in live gene banks at – Central Plantation Crops Research Institute (CPCRI), Kasaragod and the World Coconut Germplasm Centre, Andamans and sub samples are being maintained in the State Agricultural Universities. The Coconut germplasm repository at the CPCRI consists of 86 exotic and 46 indigenous accessions including 24 exotic accessions collected under the FAO/IPGRI funded expedition in 1981-82 which are being maintained at the World Coconut Germplasm Centre, Anadamans. The exotic collection consists of 68 Talls, 16 dwarfs, one semitall and one hybrid from 22 countries. The indigenous collection comprises of 34 talls and 12 dwarfs.

Sub samples of the germplasm are also being maintained at Pilicode under the Kerala Agricultural University and 10 Co-ordinating Centres under the AICRP on Palms in the State Agricultural Universities (Table 1.5 (Nair *et.al.*, 1996).)

Table 1.5 : Coconut germplasm holdings at different Co-ordinating Centres

S.No.	Centre	Exotic		Indigenous		Total
		Tall	Dwarf	Tall	Dwarf	
1.	Aliyarnagar	20	3	19	5	47
2.	Ambajipeta	6	–	5	–	11
3.	Arsikere	8	2	1	–	11
4.	Jagadalpur	7	2	1	–	10
5.	Kohikurichi	5	2	3	1	11
6.	Jalalgarh	7	2	1	–	10
7.	Konarak	9	3	1	3	16
8.	Mondouri	9	3	1	–	13
9.	Ratnagiri	14	–	12	2	28
10.	Veppankulam	12	3	8	5	28

The performance of promising varieties of coconut at CPCRI, Kasaragod is shown in Table 1.6.

Table 1.6 : Performance of promising exotic and indigenous cultivars of coconut at Kasaragod

Cultivar	Mean annual yield (nuts/palm)		Copra	
	9th and 10th year	17th-20th year	Per nut (g)	Per palm (kg/year)
Exotic				
Fiji Tall	37.4	106.1	180	19.1
Fiji Long Tong Wan	19.9	104.3	181	18.9
Philippine Ordinary	37.3	107.7	210	22.6
Philippine Laguna	25.3	87.5	261	22.8
Strait Settlement Green	46.8	108.1	196	21.2
Indigenous				
Kappadam	18.5	90.0	299	26.9
Andaman Ordinary	50.3	94.0	168	15.8
Lakshadweep Ordinary	57.3	98.3	172	16.9
West Coast Tall	22.3	68.2	176	12.0

Philippines ordinary among the exotics and Kappadam in indigenous collections have performed well at CPCRI Kasaragod.

A study on the characterization of coconut germplasm consisting of 62 exotic and 32 indigenous accessions was made by Rao and Pillai (1984) based on fruit component analysis. The accessions were divided into four major geographical sets based on the location of collection, namely, accessions from countries in South-East Asia, Pacific Islands, North Central and South American and African region. Data on yield and nut characters in 40 accession were reported. The annual out-turn of copra ranged from 7 to 30 kg per palm in Tall varieties while in the Dwarfs it was less than 10 kg per palm.

The high values for the total out-turn of copra (copra yield kg/palm/year) of Kappadam (26.9 kg), San – Ramon (22.1 kg), Java (30 kg), Philippines Laguna (19.1 kg) and Federated Malayan States (19.7 kg) confirm the contention that these are evolved as a result of selection pressure operating towards higher copra.

Phenotypic stability for annual yield of nuts of 32 coconut cultivars grown at Regional Agricultural Research Station, Pilicode was estimated by non-parametric procedures using 20 years yield data from 1965 to 1984. The non-parametric measures of phenotypic stability viz., mean of the absolute rank differences of the cultivar over the environments and variance among the ranks over the environments, were used for analysis. There was significant differences in the stability of the cultivars. West Coast Tall, Lakshadweep Ordinary, Java, Fiji, Philippines Ordinary, Andaman Ordinary and Godavari were found to be stable cultivars. Cochin China, Strait Settlement Green and Lakshadweep Small were highly unstable.

Since some cultivars showed biennial (alternate) bearing tendency, the analysis was conducted using the cumulative yield of two consecutive years also. West Coast Tall, Lakshadweep Ordinary, Andaman Ordinary, Philippines, and Java were stable for annual yield whereas Kappadam and Cochin China, were stable for cumulative yield of two consecutive years only (Balakrishnan *et. al.*, 1988).

Metroglyph analysis was done in a group of 23 genotypes of coconut maintained at Coconut Research Station (CRS), Veppankulam. The varieties came under two groups based on morphological characters. Three groups were recognised when the classification was made on the basis of nut characters. The exotic cultivars, in general, had relatively higher expression for morphological and nut characters. The possibility of obtaining

Table 1.7 : Germplasm Set I (33 and 35 years old trees) Annual Yield (nut and copra) and nut characters (1996-97) at Veppankulam

S.No.	Genotypes	Whole nut wt (g)	Dehusked nut wt (g)	Kernel wt/nut (g)	Copra wt/nut (g)	Annual yield Nuts/ palm	Annual yield Copra/ kg/palm
	Talls						
1.	WCT	541	310	175	125	221	12.2
2.	LO	880	410	195	140	167	8.8
3.	CC	1420	721	360	170	150	9.0
4.	NG	1410	850	385	165	89	4.9
5.	AO	1250	631	300	170	152	12.1
6.	FMS	1350	751	380	171	128	10.5
7.	Ph.O	1310	810	390	165	133	9.0
8.	JG	1305	800	365	161	129	8.0
9.	Guam	660	360	260	139	139	6.7
10.	AG	1420	850	365	187	103	8.5
11.	Kappadam	1320	690	297	151	102	8.3
12.	Siam	1210	711	311	155	135	8.6
13.	SSG	810	450	195	140	111	8.9
14.	LM	540	230	170	115	201	10.3
15.	Fiji	850	425	280	159	111	10.1
16.	JT	1575	890	431	210	45	6.6
17.	Sanblas	1500	850	401	186	61	3.6
18.	Spicata	561	350	185	130	182	10.9
19.	Thailand	854	595	265	142	119	8.4
20.	ECT	710	391	190	140	198	8.4
	Dwarfs						
1.	MRD	721	471	222	134	83	4.4
2.	COD	698	489	217	120	97	4.9
3.	MYD	770	477	219	111	60	5.6
4.	MGD	1103	655	268	120	103	6.6
5.	AY	487	379	156	90	130	5.1
6.	GA	989	554	260	136	88	8.2
7.	AOD	751	402	175	126	28	4.8
8.	CGD	391	198	93	60	38	2.4

WCT : West Coast Tall, LO : Lakshadweep Ordinary, CC : Cochin China, NG : New Guinea, AO : Andaman Ordinary, FMS : Federated Malay States, Ph.O : Philippines Ordinary, JG : Java Giant, AG : Andaman Giant, SSG : Strait Settlement Green, LM : Lakshadweep Micro, JT : Java Tall, ECT : East Coast Tall, MRD : Malaysian Red Dwarf, COD : Chowghat Orange Dwarf, MYD : Malaysian Yellow Dwarf, AY : Ayiramkachi, GA : Gangabondan, AOD : Andaman Orange Dwarf, CGD : Chowghat Green Dwarf.

heterotic hybrids from divergent parental combinations was indicated (Raveendran *et.al.*, 1987).

Yield evaluation made on 20 Tall and 8 Dwarf cultivars at CRS, Veppankulam (33 and 35 years) in the 1st set of planting is presented in Table 1.7.

Among the Talls, superior performance was observed in WCT with an annual yield of 221 nuts, and 12.2 kg of Copra per palm. Highest values were observed in Java Tall for whole nut weight (1575 g), dehusked nut weight (890 g), kernel wt (431 g), and Copra wt (210 g) per nut. However, due to low annuai nut yield (45 nuts) the copra yield (6.6 kg) was low in the variety. Among the Dwarfs, MGD was superior in whole nut weight and dehusked nut weight, while the high copra weight (136 g) per nut resulted in high annual copra yield of 8.2 kg per palm in Gangabondan.

Table 1.7a : Germplasm Set II (13 years old trees under planted) Annual yield (nut and copra) and nut characters (1996-97) at Veppankulam

S.No.	Genotypes	Whole nut wt (g)	Dehusked nut wt (g)	Kernel wt/nut (g)	Copra wt/nut (g)	Annual yield	
						Nuts/ palm	Copra/ kg/palm
1.	NT	821	440	256	130	50	2.9
2.	San Ramon	1710	1010	389	217	32	3.7
3.	WCT	810	425	201	118	77	3.8
4.	CC	815	440	231	111	29	1,5
5.	BSI	820	430	260	161	47	2.9
6.	Kenya	702	405	217	95	55	3.1
7.	Guam	610	325	165	95	47	1.7
8.	Zanzibar	825	551	246	176	61	7.7
9.	Seychilles	749	450	279	125	73	4.2
10.	St. Vincent	890	475	225	131	44	2.5
11.	ECT	790	435	220	125	76	3.8
12.	MYD	630	315	155	75	37	1.1

NT : Natal tall, WCT : West Coast tall, CC : Cochin China, BSI : British Soloman Islands, ECT : East Coast tall, MYD : Malaysian Yellow Dwarf.

A second set of germplasm consisting of nine exotic Tall plus one dwarf cultivars of 13 years age were compared with two local tall cultivars of ECT and WCT of the same age group under a situation of under planting. San Ramon was found to score highest values for whole nut weight, dehusked nut weight, kernel weight and copra weight per nut (217 g). However, due to its low annual nut yield (32) the copra yield was comparable to WCT which gave 77 nuts/palm. With fairly high copra weight per nut (176 g) and moderate nut yield (61) the copra yield per palm was the highest (7.7 kg) in Zanzibar, compared to the rest of the varieties in relative terms.

Germplasm Evaluation at different Centres of All India Co-ordinated Research Project on Palms

At Ambajipet (Andhra Pradesh), Lakshadweep Ordinary, Philippines Ordinary, Fiji, Cochin China and Java are the high yielders compared to East Coast Tall. The average nut yield ranges from 96.3 to 118.2 per palm/year. The evaluation of germplasm have resulted in the identification of high yielding cultivars\varieties. Out of the 62 cultivars evaluated at Pilicode (Kerala), 9 cultivars have yielded above 2.6 tonnes of copra per hectare under rainfed conditions as shown in Table 1.8.

Table 1.8 : Performance of Coconut Cultivars under rainfed conditions at Pilicode

S. No.	Cultivar	Mean nut yield per palm per year	Copra content per unit (g)	Copra yield Kg/palm	Copra yield Kg/ha
1.	Spicata	116.2	175.0	20.3	3617
2.	Philippines Ordinary	79.3	236.5	18.8	3339
3.	New Guinea	75.2	247.3	18.6	3309
4.	Lakshadweep Ordinary	93.1	193.7	18.0	3211
5.	San Ramon	51.3	346.0	17.8	3159
6.	Cochin China	67.3	243.9	15.8	2814
7.	Java	66.1	236.3	15.6	2778
8.	Andaman Giant	78.7	189.2	14.9	2650
9.	Siam	78.8	189.0	14.9	2652
10.	West Coast Tall	67.5	180.0	12.2	2126

Source : Rethinam (1989).

Of the 30 cultivars with a bearing period of 20.5 to 22.5 years evaluated at Veppankulam (Tamil Nadu), seven cultivars have surpassed the East Coast Tall in terms of nut and copra yield (Table 1.9).

Table 1.9 : Nut and Copra yield of Coconut Cultivars at Veppankulam

S. No.	Cultivar	Mean nut yield per palm per year	Copra content per nut (g)	Annual Copra Yield per palm (kg.)	Annual Copra Yield per ha. (kg.)
1.	Malaysian Green Dwarf	93.2	153	14.3	2495
2.	Andaman Giant	67.1	268	18.0	3147
3.	Federated Malay States	72.6	179	13.0	2274
4.	Fiji	80.4	144	11.6	2026
5.	Philippines	59.7	153	9.1	1590
6.	Andaman Ordinary	76.6	193	14.8	2587
7.	Cochin China	58.0	196	11.4	1989
8.	East Coast Tall	74.5	115	8.6	1499

The performance of germplasm at Ratnagiri revealed the superiority of Lakshadweep Ordinary and Banawali Green Round.

Table 1.10 : Yield performance of germplasm accessions at Ratnagiri

S. No.	Cultivar	Mean annual nut yield/palm (1979-88)	Oil Yield Kg/ha/yr
1.	Lakshadweep Ordinary	154	2808
2.	Banawali Green Round	148	3013
3.	Banawali Yellow Round	131	–
4.	Federated Malay States	116	1701
5.	Banawali Green Long	112	2312
6.	Fiji	111	2284
7.	Banawali Yellow Long	131	2128
8.	Banawali Red Round	108	1961
9.	Philippines Ordinary	107	2794
10.	New Guinea	102	2098
11.	West Coast Tall	94	2691

Source : Rethinam (1989).

4. Hybridization

Harland (1957) has indicated that it is possible by controlled pollination to make hybrids where both the mother and father are known, designated as *'paired crosses'*, which formed the basis of a method of tree breeding as some of the high yielding paired crosses have given a yield more than twice the average; resowing of the proved high yielding pairs resulted in a gain of 30 per cent over the unselected commercial material and further cycle of paired crosses between the best of the paired crosses have given still further increase in yield so that with proper management, it reached a level about double that of the original unselected type. Harland has also indicated the necessity of determining the prepotent material transmitters which inspite of having been indiscriminately pollinated by miscellaneous males are sufficiently possessed of dominant yield factors to ensure that their offsprings are also high yielding. He further opined that once a material transmitter identified, it can continuously be used in paired crosses, self fertilization and as male paired in extensive crosses with high yielding mother palms. This method, according to him, is the most immediately practical one which can be adopted. He also observed that the process of identifying male transmitters can be speeded up by the use of dwarf palms as females since they are largely self pollinated and reasonably homoegenous. He has suggested using a number of dwarfs as female parents and pollinating each bunch of different high yielding Tall male parent and studying their progeny to identify the most prepotent males. In studying their progenies, he suggested planting 5-10 seedlings from each cross to make a tentative separation.

Results of hybridization work carried out in different countries are presented below.

Philippines

The hybrids (Dwarf × Tall) evolved and recommended for planting are PCA 15-1 (Catigan dwarf × Laguna tall), PCA 15-2 (Malayan red dwarf × Tagnanan) and PCA 15-3 (Malayan red dwarf × Baybay). The use of these (D × T) hybrids as well as the local Baybay cultivar has been found to be more profitable investment than that of MAWA hybrid and other local talls (Eleazar, 1996).

Indonesia

The Coconut Research Institute (CRI) has released three types of Indonesian Coconut hybrids, namely Khina-1, Khina-2 and Khina-3. The materials used for these Dwarf × Tall hybrids were indigenous. In Khina-1, the mother plant is Nias Yellow Dwarf while the pollen parent is Tenga Tall. In Khina-2 and Khina-3, the mother plant is same (i.e.) Niyas Yellow Dwarf, while the pollen parents are Bali Tall and Palu Tall respectively. The fruit is green in all the three hybrids and the fat content varies from 60.6 to 62.5 and copra production is around 4 t/ha. All the three are D × T hybrids.

In addition to the three D × T hybrids, CRI has also released four T × T hybrids by crossing selected tall cultivars in Kima Atas Experimental Garden. These improved talls are called "Kelapu Baru" (KB). They are KB-1, KB-2, KB-3 and KB-4. KB-1 is a cross between cultivars No. 32 and No. 33. The fruit is reddish brown, fat content 68 per cent and copra production is 3.9 t/ha. KB-2 is result of crossing between cultivars No. 32 and No. 2, the fruit is greenish yellow, fat content 68 per cent and copra production is 4.5 t/ha. KB-3 is a cross between cultivars No. 32 and 83, the fruit is reddish brown, fat content 71 per cent and copra production is 4.7 t/ha. KB-4 is an outcome of cross between cultivars No. 32 and No. 99, the fruit is reddish yellow, fat content 67 per cent and copra production is 4.1 t/ha. The fat content in copra as well as copra output of the tall × tall hybrids are higher than that of dwarf × tall hybrid (Michellia Darwis, 1966).

Dwarf × Tall

Four varieties/cultivars suitable for use in the breeding programme were identified from the germplasm survey carried out in 11 provinces of Indonesia. They were, Nias Yellow Dwarf (NYD), Tanga Tall (TT), Palu Tall (PT) and Bali Tall (BT). NYD was crossed with the three Tall forms to study the performance of the first generation progenies. Simultaneously seed gardens were established using NYD as the female parent and the three Tall forms as the male. They were capable of producing NYD × TT, NYD × BT and NYD × PT seed. The seed gardens were developed according to a new concept.

In 1982, the F_1 palms of NYD × TT, NYD × BT and NYD × PT have proved their superiority to the Ordinary Tall in respect of precocity for bearing and high copra production. In 1983 over two million seednuts of improved varieties were available.

Seven commercial plantations were set up from 1967 to 1978 and 75% of the 16,500 ha covered were planted with Dwarf × Tall hybrids. (90% under PB 121 and 10% under PB 111 and PB 141) and 25% Local West African Talls (WAT). Comparisons were made between the hybids and WAT for annual yields obtained during 1976 to 1990 and resistance to *Phytopthora heveae* and *Eriophyes guerreronis*. The relationsip between yield, climate and soil were also determined. The hybrids were found to be superior to WAT with regard to precocity, number of nuts per hectare and copra per nut (on average 80% higher) even though after 17 years they have not reached their production ceiling. Variations in the best yields were obtained on deep soils, Hybrids were more tolerant to *E. guerreronis* but more susceptible to nut-fall caused by *P. heveae* than WAT.

France

In 1942, during Second World War, the Institute de Recherchespourles Huiles et Oleagineux (IRHO) was set up as non-profit making organisation (French law of 1901) to develop oil crops (annual and perennial), primarily in Africa. Priority was initially given to oilpalm, but coconut and groundnut were far from overlooked. The first coconut station was set up in 1949 at Seme Podji (Benin), followed in 1950 by Port Bouet (Ivory Coast), which became the Marc Delorme Station in memory of one of its Directors. Seme Podji initially covered 200 ha and Port Bouet 450, later on increasing to 1000 ha. Besides Africa, IRHO extended its activities to other continents : Rangiroa in French Polynesia, Saraouton (Santo) in Vanuatu, Cambodia, Brazil, French Guiana, Indonesia and the Philippines.

The first hybrids produced, tested and extended were Dwarf × Tall ecotypes which exhibited desirable traits such as precocity in flowering and higher yields. These hybrids recorded a cumulative yield of 7.5 t/ha of copra at 6 years compared to 0.9 t/ha in tall types grown under the same conditions. When fully grown, the hybrids produce 1.5 to 2.5 times more than talls. Some Tall × Tall hybrids are also promising and facilitate intercropping or livestock rearing under them. In addition to developing the hybridization programme, IRHO rapidly developed a simple, reliable and cheap way of reproducing the hybrids in seed gardens and extending them.

The PB 121 hybrid (MYD × WAT) was initially the "Star" and is the most widely planted hybrid in the world. It has proved remarkable mainly due to the excellent compatibility between the parents–Malaysian Yellow Dwarf (MYD) and West African Tall (WAT). It shows significant heterosis for number of fruits and copra per nut.

Other Dwarf x Tall hybrids are also promising, primarily the Cameroon Red Dwarf (CRD) x West African Tall (WAT), Malaysian Yellow Dwarf (MYD) x Rennel Island Tall (RIT) and Malaysian Red Dwarf (MRD) x West African Tall (WAT). The Tall x Tall hyrbids such as the West African Tall x Rennel Island Tall are equally efficient in the production of nuts and copra per palm (Rognon, 1996).

Table 1.11 : Characteristics of high yielding hybrids at the Marc Delorme Station (Rognon, 1996)

Characteristics	Hybrid Combinations			
	Dwarf x Tall			Tall x Tall
	PB 121 (MYD x WAT) (MAWA)	PB 111 (CRD x WAT)	PB 132 (MRD x WAT)	PB 212 (WAT x RLT)
Preflowering period (months)	54	44	50	51
No. of bunches per year	14.8	15.2	14.2	12.7
No. of nuts per tree	109	126	110	102
Copra contents per nut(g)	247	240	253	311
Copra yield t/ha	3.67	4.18	3.80	4.26

Sri Lanka

Early studies on coconut breeding at the CRI (Coconut Research Institute) were basically planned to identify elite palms in already existing population, crossing them and evaluating their progenies. Elite palm were selected on the basis of 100 nuts per palm and 220 g copra per nut (22 kg copra per palm) under average environmental conditions on heritiability estimates of these economic characters. Two series of cross were effected. The first series include Tall x Tall crosses of selected Sri Lanka Talls (SLT) and Tall x Dwarf. The Dwarf male parent comprises Green Dwarf (Pumila) palms. The yield performance from 5th to 12th year after planting is summarised in Table 1.12.

Table 1.12 : Yield performance of Tall x Tall and Tall x Dwarf hybrids

Parental combination	Mean annual yield (5th to 12th year after planting) (Based on 158 palms/ha)			
	Nuts/ palm	Copra kg/palm	Nuts/ha	Copra kg/ha
SLT x SLT	74	17.62	11725	2750
SLT x GD	91	21.96	14376	3470

Source : Liyanage *et.al.*, 1988.

In the second series of crosses, three sets of Tall x Dwarf and one set of Tall x Tall crosses were made. The Dwarf male parents and Green Dwarf (Pumila), Yellow Dearf (edurna), and Red Dwarf (regia). The Tall parent comparises selected Sri Lanka Tall (typica) palms. The yield data of these hybrids are furnished in Table 1.13.

Table 1.13 : Yield performance of T x D and T x T hyrbids

Parental combination	Mean preflowering period (months)	Mean annual yield (9 to 12 year old) (Based on 158 palms/ha)			
		Nuts/ palm	Copra kg/palm	Nuts/ha	Copra kg/ha
SLT x GD	34.8	64	11.47	10,110	1810
SLT x YD	37.1	67	11.61	10,590	1830
SLT x RD	38.9	55	9.58	8,690	1510
SLT x SLT	52.9	39	7.55	6,160	1190

Source : Liyanage *et.al.*, 1988.

It is clear from the above data that Tall x Dwarf hybrids outyielded Tall x Tall hybrids.

As a result of the hybridization programme between selected mother palms and prepotents, two tyeps of improved cultivars, Tall x Tall (CRIC-60) and Dwarf x Tall (CRIC-65) were released for planting in the sixties. These two hybrids have out-performed the ordinary Tall (open pollinated) variety palms used earlier for planting, in terms of nuts as well copra production. (Liyanage, 1966).

The details of the improved coconut varieties released are as under :

CRIC-60

Derived by crossing selected SLT x SLT palms. Late flowering (55 to 70 months), outbreeding, hardy palms tolerating a wide range of environmental conditions, capable of producing over 110 nuts per palm per year with about 225 g copra per nut under rainfed condition with a satisfactory environment combined with good management. Released to the industry in 1960. Recommended for planting in all the coconut growing areas of Sri Lanka, where conditons are suitable for coconut cultivation as it has got wide adapatbility.

CRIC-65

Derived by crossing GD x SLT involving Green Dwaf (Pumila) x Sri Lanka Tall palms. Early flowering (36 to 45 months), outbreeding, tolerates only a restricted environment, capable of producing over 120 nuts/palm/year with about 210 g copra per nut. Released to the industry in 1965. Recommended for planting in the home gardens or under irrigated conditions.

Other Crosses

King coconut (KC) x Green Dwarf (GD) : The KC female parent has orange red nuts and GD – green nuts. Their F_1 palms have reddish brown nuts. The nuts of the hybrids are small in size, giving about 100 nuts per palm per year with 160 g copra per nut. They are less hardier than SLT palms and are more susceptible to pest damage. They are unsuitable for cultivation on commerical scale (Liyanage, 1957).

San Ramon x GD : 34 progenies have a mean pre-flowering period of 50.6 months, yielding 68 nuts in the 9th year with 250g copra per nut.

Jamica

Results of many years of observation on the performance of F_1 hybrid coconuts in Jamica show that the F_1 hybrids have characteristics which are intermediate between those of their parents, and indicate that Dwarf x Tall hybrids hold the greatest promise for future plantings. In view of the lethal yellowing disease having reached devastating proportions in Jamaica, the hybrids to be used should involve the highly resistant Malayan Dwarf and individually surviving palms from those tall varieties identified as having an intermediate level of resistance. (Harries, 1974).

The common coconut variety in Jamaica (Jamaica Tall) is susceptible to lethal yellowing disease. The disease resistant variety Malayan Dwarf shows a high production under good circumstances, but produces small nuts on poor soil and after dry spells. The MAYPAN, a hybrid from Malayan Dwarf x Panama Tall, has bigger nuts and satisfactory disease resistance and is recommended for commercial planting. (Harries and Romney, 1974).

India

Coconut hybridization work in India since 1930 was reviewed by Satyabalan and Nambiar (1968). The early work consisted of crosses between Tall varieties possessing good economic characters, reciporcal crosses between parents with high female-flower production and high seed setting, crosses between regular bearers and high yielding irregular bearers and intervarietal crosses between Tall (female) and Dwarf varieties. It was shown that cross pollination and seedlings from high yielding groups are more vigorous than those from low, especially those producing fewer than 40 nuts per tree per year. As intervarietal crosses are promising, several were made using exotic types, one of the best being Lakshadweep Ordinary x Gangabondam, T x D, T x Gangabondam and D x D hybrids were found to be early bearing.

The importance of utilising the preococious bearing character of Dwarfs in breeding was first realised by Patel (1937) who observed maximum hybria vigour when the Tall variety was used as female parent with Dwarf as the male.

The expression of hybrid vigour in coconut palms in the components of the fruit was studied at Kasaragod (Kerala) in three hybrid progenies having Westcoast Tall (WCT) as the female parent and Dwarf Green (DG), Dwarf Orange (DO) and Gangabondam (GA) as their respective male parents. Weights of nut water, of kernel and wet weight of shell, copra content exceeded those of the superior parent in WCT x DO and WCT x GA, and exceeded tha mean values of the two parents in WCT x DG. The weight of the unhusked nut was always inferior to the mean value of the two parents, showing that all hybrid progenies form a mush thinner husk than the parent with the thickest husk (Satyabalan *et.al.*, 1970).

As an outcome of research work done at Coconut Research Station, Veppankulam in Tamil Nadu, two high yielding hyrbids VHC_1 and VHC_2 were developed and released for large scale cultivation in Tamil Nadu. VHC_1 (Veppankulam hybrid Coconut-1) is a cross between selected East Coast Tall as female parent and Malaysian Dwarf Green as male parent and released in 1982 whereas VHC_2 (Veppankulam Hybrid Coconut-2) is a product of hybridization between choice East Coast Tall palms (as female) and Malaysian Dwarf Yellow (as male) which was released in 1988 (Ramanathan *et.al.*, 1984, Ramachandran *et.al.*, 1990).

The yield performance and nut component data are furnished in Tables 1.14 and 1.15.

Table 1.14 : Yield performance of VHC_1 and VHC_2 hybrids

Particulars	T x D Hybrids		ECT
	VHC_1	VHC_2	
Number of nuts/ha	17000	18250	10830
Copra yield t/ha	2.70	2.90	1.35
No. of nuts required to make one tonne of copra	6305	6289	8000
Oil yield t/ha	1.85	2.03	0.86
Oil yield as percent on ECT	215.1	236.1	100.0

Table 1.15 : Nut component and Oil content

Hybrid/ Check	Whole nut weight (g)	Dehusked nut weight (g)	Kernel weight (g)/nut	Copra weight (g)/nut	Oil content percent	Annual copra yield Kg/palm
VHC_1	992 ± 54	602 ± 30	280 ± 20	158 ± 4	68.6	20.7
VHC_2	1002 ± 79	615 ± 18	314 ± 14	159 ± 78	70.2	23.3
ECT	945 ± 106	518 ± 58	253 ± 47	125 ± 7	63.5	10.3

Fig. 1.1 : Malaysian Dwarf Green (MDG) Parent

Fig. 1.2 : VHC_1 Coconut hybrid (ECT × MDG)

Fig. 1.3 : VHC_2 Coconut hybrid (ECT × MDY)

Fig. 1.4 : VHC_2 Coconut hybrid (Five years old)

Fig. 1.4a : Malaysian Dwarf Yellow (MDY) Parent

Fig. 1.5 : Fruit bunches. ECT, VHC_2 – MDY

The copra yield is more than double in VHC_1 and VHC_2 compared to ECT, while in oil yield VHC_1 has shown 115 per cent increase and VHC_2 excelled by 136 per cent over and above that, of East Coast Tall (ECT).

The hybrid vigour has been successfully exploited on a commercial scale. The hybrids developed in three states viz.,

Kerala, Tamil Nadu and Andhra Pradesh involving talls and dwarfs as parents are not only early bearing but also give higher yield than either of the parents as shown in Table 1.16 based on the performance reported at CPCRI, Kasaragod (Kerala). (Nair *et. al.*, 1996).

From a study conducted at the Coconut Research Station, Veppankulam, on the incidence of grey blight disease in several varieties and hybrids, it was clear that East Coast Tall × Malaysian Dwarf Green has shown fairly low disease index of 17.0 as against 21.9 per cent in East Coast Tall. The disease score is intermediate between tall parent (21.9) and dwarf parent (10.2) (Bhaskaran and Ramanathan, 1983).

Table 1.16 : Performance of released Coconut hybrids

Hybrids/Control	Annual nut yield/ palm	Copra content g/nut	Copra yield/yr. Per palm (kg)	Copra yield/yr. Per ha (t)	Oil Content percent	State for which recom-mended
D × T Chandrasankara (COD × WCT)	116	215	25	4.4	68	Kerala, Karnataka, Tamil Nadu
T × D Kerasankara (WCT × COD)	108	187	21	3.5	68	Kerala Karnataka
T × D Chandralaksha (LO × COD)	109	195	21	3.7	69	Kerala Karnataka
T × D Lakshaganga (LO × GB)	108	195	21	3.7	70	Kerala
T × D Anandaganga (AO × GB)	95	216	21	3.6	68	Kerala
T × D Keraganga (WCT × GB)	100	201	21	3.5	69	Kerala
T × D VHC_1 (ECT × MDG)	98	135	13	2.3	70	Tamil Nadu
T × D VHC_2 (ECT × MDY)	107	152	16	2.9	69	Tamil Nadu
T × D EOT × GB	140	150	21	3.7	68	Andhra Pradesh
WCT (Control)	80	176	14	2.5	68	Control

COD : Chowghat Orange Dwarf, LO : Lakshadweep Ordinary, AO : Andaman Ordinary, MDG : Malaysian Dwarf Green, MDY : Malaysian Dwarf Yellow, GB : Gangabondam, ECT : East Coast Tall, WCT : West Coast Tall.

VHC_3 Coconut Hybrid

A Coconut hybrid developed at the Coconut Research Station Veppankulam (Tamil Nadu) utilising a well adopted "East Coast

Tall" (ECT) as female parent and "Malaysian Orange Dwarf" as male parent was released during January 2000, as VHC_3 for large scale cultivation in Tamil Nadu. With a nut weight of 908 g./nut and mean annual yield of 156 nuts/Palm it is superior to VHC_2 by 10 per cent. The Kernel and Copra weight are 297 g and 162 g. respectively per nut. The copra yield per hectare is 3.65 tonnes as against 1.75 tonnes of "East Coast Tall" and it is 19.7% higher than that of VHC_2. One tonne of Copra is extracted from 6180, 6675 and 7680 nuts of VHC_3, VHC_2 and ECT, respectively. The estimated oil yield per hectare of VHC_3 hybrid is 2.55 tonnes as against 2.13 and 1.13 tonnes of VHC_2 and ECT respectively. VHC_3 hybrid performs better under better management condition.

Exploitation of Heterosis

Four sets of intervarietal hybrids involving East Coast Tall as female parent (ECT) and Ayiramkachi (AY), Malaysian Dwarf Green (MDG), Malaysian Dwarf Yellow (MDY) as male parents and a reciprocal hybrid involving Malaysian Dwarf Yellow as female and ECT as male were studied at Coconut Research Station, Veppankulam. The results have shown that ECT × AY hybrids have excelled their parents in nut yield, and weight of dehusked nuts. ECT × MDG hybrids have excelled their parents in nut, and copra yield, oil cotent and oil yield. ECT × MDY has also shown similar performance except in oil content which is comparable to ECT. The MDY × ECT hybrids (D × T) have excelled their parents in weight of dehusked nuts, copra and oil yield per hectare (Ramanathan and Louis, 1983).

In four sets of hybrids, nut and copra yield, nut components and oil content were compared among hybrids and parents. The hyrbids have exhibited different degrees of heterosis for the characters studied. VHC_1 hybrid was found to express the highest vigour exceeding the values of greater parent and also other hybrids in terms of nut and copra yield, nut components and oil content (Ramanathan, 1989a).

Oil content was studied in twelve sets of hybrid combinations and their parents. The highest expression of heterosis over the mid-parent value was evident in East Coast Tall × Gangabondam (13.4%) and East Coast Tall × Kanyakumari Dwarf Yellow (T × D) (13.4%) closely followed by Kanyakumari Dwarf Yellow × East Coast Tall (D × T) with 12.7 per cent. The percentage of heterosis over the better parent in respect of East Coast Tall (ECT) × Gangabondam (GB) and East Coast Tall × Kanyakumari Dwarf Yellow (KDY) were 12.4 each, while in KDY × ECT it was 11.7. Thus, there is difference in the reciprocal crosses in the expression of heterosis for oil content. This was also clear in Gangabondam (GB) × East Coast Tall (ECT) hybrids with 7.1 per cent hetrosis for oil content as against 13.4 per cent in its reciprocal cross hybrid (Nallathambi *et. al*, 1986).

Performance of Hybrids in Dry Conditions

The performance of two sets of hybrids, East Coast Tall (ECT) × Malaysian Dwarf Green (MDG) and East Coast Tall (ECT) × Chowghat Dwarf Orange (CDO) were compared with ECT under unirrigated dry conditions. Results have shown that ECT × MDG was significantly superior to ECT × CDO and ECT in nut yield. The annual nut yield per palm was 76, in ECT × MDG as against 32 and 29 in ECT × CDO and ECT respectively (Ramanathan and Nallathambi, 1988).

Evaluation of yield performances of 87 hybrid combinations in coconut indicated the superiority of CDO × WCT, LO × GB and LO × CDO under rainfed conditions over WCT. CDO × WCT gave the highest cumulative yield (46% higher than WCT) at the end of 18th year. LO × GB and LO × CDO in addition to their higher yield potential over the WCT, also gave increased yields (140 and 84 per cent higher than WCT) even during drought affected years (Anon, 1985).

An experiment was conducted at RARS, Pilicode, (Kerala) to study the best hybrid coconut under rainfed conditions among 15 hybrid combinations planted during 1973. Yield performance of these hybrids upto 1988 have shown that two hybrids, West Coast Tall (WCT) × Malaysian Dwarf Yellow (MDY) and West Coast Tall (WCT) × Chowghat Dwarf Green (CDG) were superior in nut, and copra yield followed by WCT × CDO and Fiji × CDO (Balakrishnan and Kannan, 1988).

Eighty seven coconut hybrids comprising of 21 different Tall × Dwarf, 10 Dwarf × Tall, 51 Tall × Talls, 4 Dwarf × Dwarfs and one three way hybrid were studied. Evaluation on 18th year based on cumulative yield which was representative of true yield potential, was carried out and three hybrid combinations, namely, Chandrasankara (CDO × WCT), Chandralaksha (LO × CDO) and Lakshaganga (LO × GB) were identified as superior hybrid combinations and released for Kerala. These hybrid combinations gave 19 to 42 per cent higher yield than cultivated WCT. Among these three hybrids, Chandrasankara (CDO × WCT) was found to give 47% higher copra than WCT. Hybrids, Lakshaganga (LO × GB) and Chandralaksha (LO × CDO) performed well under drought situations. The drought tolerance in these hybrids was identified to be the contribution from the female parent Lakshadweep Ordinary (LO). (Rao *et. al.*, 1988).

Seven varieties and three hybrids were studied under semidry conditions at the Department of Oil Seeds, Tamil Nadu Agricultural University, Coimbatore. Nut and copra yield and nut components were compared with each other. The hybrids WCT × MDY and its reciprocal MDY × WCT were found to be superior with relatively higher annual nut yield per palm. Among the varieties, Lakshadweep Micro performed better in terms of nut and copra yield and nut components (Ramanathan *et al.*, 1992).

Physiological Basis for Drought Tolerance

Twnety three genotypes studied in Kerala had shown varying response to the prevailing soil and atmospheric conditions during summer months. Tall genotypes had a relatively high stomatal resistance (RS), leaf water potential and epicuticular wax content as compared to the dwarfs. Lower transpiration rate resulting in effective conservation of water in tissues was observed between November and March in some tall palms and hybrids. Parametric relationship with drought tolerance based on the rank sum led to the identification of few genotypes with the desirable drought tolerance characters (Rajagopal *et.al.*, 1990).

The genotypes COD × WCT, its reciprocal hybrid WCT × COD and WCT responded to moisture stress to different degrees. WCT × COD and WCT were characterised by higher stomatal resistance and higher leaf water potential. Epicuticular wax content as well as acid phosphatase (APH) and glutamate oxaloacetate transminase (GOT) activities increased under stress. The total nut yield and kernel dry weight decreased in COD × WCT than in other two genotypes under both rainfed and irrigated conditions (Rajagopal *et.al.*, 1988(a)).

Adult coconut palms comprising the talls, West Coast Tall (WCT) and Lakshadweep Ordinary (LO), the dwarfs, Chowghat Orange Dwarf (COD) and Gangabondam (GB) and the hybrids (WCT × COD), COD × WCT, LO × GB and LO × COD) were screened for drought tolerance using leaf water potential (LWP) measurement. When subjected to dehydration the LWP declined with time to different degrees among genotypes except for LO × COD, LO × GB and WCT which had higher LWP irrespective of time of dehydration. These three genotypes had lower activities of acid phosphatase (APH) and glutamate oxaloacetate transminase (GOT). Thus, LWP measurement serves as an index of drought tolerance in coconut (Rajagopal *et.al.*, 1988(b)).

Coconut genotypes were screened to assess the degree of tolerance to stress based on the soil moisture depletion by roots, transpiration rate, leaf surface wax (ECW) and enzyme activities. Thus, the hybrids LO × GB ×, LO × COD, WCT × COD and WCT × WCT and the talls WCT and Andaman Giant (AG) have proved to be drought tolerant in having lower soil moisture depletion, lower transpiration rate and higher ECW as compared to the dwarfs like COD, MYD, GB, MOD, Lakshadweep Micro and the hybrid COD × WCT which were found to be drought succeptible. Thus there is reciprocal difference as WCT × COD was tolerant to drought. The activities of glutamate oxaloacetate transminase (GOT) and nitrate reductase (NR) also showed distinct differences between tolerant and susceptible genotypes. (Rajagopal *et.al.*, 1988(c)).

Physiological Basis for Production Trends

Attempts were made to study physiological basis for the variation in production trends in certain coconut hybrids and their parents. The hybrids in general possessed higher chlorophyll content in young leaves (14th leaf from the first unforld leaf), photosynthetic rate, soluble protein and nitrate reductase activity and low respiratory rates compared to parents. East Coast Tall (ECT) among the parents and Malaysian Dwarf Yellow (MDY) × East Coast Tall (ECT) and its reciprocal amongst the hybrids were identified as better material for the production potential in terms of high photosynthetic rates coupled with low respiratory rates. The studies indicate the possibility of using these physiological parameters in selecting vigorous seedlings in the nursery stage itself. (Raveendran *et.al.*, 1989).

Table 1.17 : Performance of Coconut hybrids (24 years old) involving Tall × Dwarf, reciprocal and Tall × Tall at Veppankulam

Sl. No.	Hybrids	Whole nut wt (g)	Dehus-ked nut wt (g)	Kernel wt\nut (g)	Copra wt/nut (g)	Annual yield	
						Nut/ palm	Copra kg/palm
1.	Siam × AY (T×D)	713	295	164	93	79	4.1
2.	AY × Siam (D×T)	583	341	201	101	67	6.1
3.	AY × CC (D×T)	840	501	231	104	120	9.4
4.	CC × AY (T×D)	788	441	219	108	105	3.3
5.	Siam × CGD (T×D)	887	448	236	102	93	5.9
6.	ECT × CGD (T×D)	546	225	154	88	103	9.2
7.	CC × MYD (T×D)	675	405	180	105	145	9.4
8.	MGD × ECT (D×T)	775	415	225	85	140	8.4
9.	ECT × MGD (T×D)	532	245	130	70	119	5.1
10.	MYD × ECT (D×T)	542	283	156	84	140	8.3
11.	ECT × GB (T×D)	630	323	203	113	151	11.4
12.	GB × ECT (D×T)	715	204	158	103	183	10.3
13.	CGD × ECT (D×T)	600	340	225	155	69	11.3
14.	AY × ECT (D×T)	540	230	145	69	86	5.9
15.	ECT × MRD (T×D)	733	420	235	121	195	14.1
16.	ECT × MYD (T×D)	565	270	163	105	224	13.9
17.	ECT × AOD (T×D)	375	110	85	38	26	1.4
18.	MYD × CC (D×T)	500	165	110	65	75	2.0
19.	CGD × Siam (D×T)	450	180	80	55	16	1.9
20.	ECT × WCT (T×T)	423	188	93	63	131	3.0
21.	ECT × AY (T×D)	550	238	126	78	80	4.6

AY : Ayiram Kaichi, MYD : Malaysian Yellow Dwarf, MGD : Malaysian Green Dwarf, MRD : Malaysian Red Dwarf, AOD : Andaman Orange Dwarf, CGD : Chowghat Green Dwarf, GB : Gangabondam, CC : Cochin China, ECT : East Coast Tall, WCT : West Coast Tall.

Varieties Suitable for Tender Coconut

Coconut genotypes numbering 79 (40 hybrids and 39 cultivars) were screened for tender coconut water trade. The identified eight genotypes in preliminary screening were further subjected to screening for volume of tender coconut water, pH, total sugars, organoleptic score for tender coconut water and tender coconut meat at 5th, 6th, 7th month of maturity. WCT in talls, COD and MOD in dwarfs are considered as the best (WCT : West Coast Tall, COD : Chowghat Orange dwarf, MOD : Malaysian Orange dwarf).

Table 1.18 : Yield Performance of Coconut hybrids (20 years old) involving Tall × Dwarf, reciprocals and Tall × Tall and Dwarf × Dwarf at Veppankulam

Sl. No.	Hybrids/Parents	Whole nut wt (g)	Dehus-ked nut wt (g)	Kernel wt\nut (g)	Copra wt/nut (g)	Annual yield	
						Nut/ palm	Copra kg/palm
1.	CC × LO (T×T)	768	465	225	140	53	10.5
2.	LO × CC (T×T)	792	468	230	145	120	12.4
3.	CC × Phil (T×T)	925	590	266	180	80	10.2
4.	Phil × CC (T×T)	835	540	275	160	103	10.3
5.	Siam × MYD (T×D)	955	655	285	195	86	16.4
6.	Siam × AY (T×D)	621	345	210	125	119	10.0
7.	ECT × AY (T×D)	710	415	230	125	111	11.0
8.	CC × MYD (T×D)	792	468	230	145	58	11.2
9.	MYD × Siam (D×T)	751	495	261	151	138	11.9
10.	AY × Siam (D×T)	614	351	205	115	71	8.5
11.	AY × WCT (D×T)	820	502	271	156	131	12.8
12.	AY × COD (D×D)	470	260	145	95	62	8.4
13.	MYD × CGD (D×D)	630	400	190	90	121	7.7
14.	CGD × MYD (D×D)	611	410	200	115	170	10.5
15.	CC (T)	890	520	280	175	117	13.8
16.	LO (T)	841	361	270	161	120	13.8
17.	Phil (T)	851	511	265	169	132	17.3
18.	Siam (T)	668	393	223	125	43	6.4
19.	AY (D)	650	350	185	85	144	8.8
20.	MYD (D)	850	451	241	131	31	7.4

AY : Ayiram Kaichi, CGD : Ch ghat Green Dwarf, MYD : Malaysian Yellow Dwarf, COD : Chowghat Orange Dwarf, LO : Lakhshadweep Ordinary, CC : Cochin China, Phil.: Philippines.

Performance of 24 years old coconut hybrids involving T × D (11 hybrids), D × T (9 hybrids) and Tall × Tall (1 hybrid) was compared and data presented in Table 1.17. Superior performance was observed in ECT × MRD among tall × dwarf hybrid, with high annual nut yield of 195 and annual copra yield of 14.1 kg per palm. Among the dwarf × talls, though the annual nut yield was modest (69) in CGD × ECT, the copra yield was the highest (11.3 kg/palm).

In another set of 20 year old hybrids involving 4, T × D combinations, three D × T, four Tall × Tall and three Dwarf × Dwarf combinations were compared with parents and the data presented in Table 1.18. High degree of heterosis was observed in Siam × MYD (T × D) in the entire set, with high values for whole nut weight, dehusked nut weight, kernel weight, copra weight per nut and annual kernel yield (16.4 kg) per palm.

Nine hybrids involving indigenous and exotic tall combinations were compared with the parents, and VHC_1 and the data presented in Table 1.19. As these were evaluated in underplanting situation, the data are not impressive.

Table 1.19 : Performance of Coconut hybrids (12 years old) involving indigenous and exotic talls at Veppankulam. (Underplanting situation)

Sl. No.	Hybrids/ Parents	Whole nut wt (g)	Dehus-ked nut wt (g)	Kernel wt\nut (g)	Copra wt/nut (g)	Annual yield Nut/ palm	Annual yield Copra kg/palm
1.	FMS × ECT	641	365	174	103	19	1.2
2.	FMS × Sanblas	583	320	188	94	28	2.1
3.	JG × ECT	503	280	210	110	21	1.5
4.	JG × Sanblas	500	220	110	55	15	0.9
5.	Phil × ECT	425	260	155	70	19	1.6
6.	Phil × Sanblas	355	170	86	46	22	1.2
7.	LO × CC	491	245	141	69	47	2.0
8.	LO × Phil	705	490	270	130	30	2.4
9.	NG × ECT	552	275	142	63	29	1.4
10.	VHC-1	603	313	180	114	43	4.1
11.	ECT	852	408	220	125	21	1.7
12.	Sanblas	840	346	273	143	32	5.6
13.	FMS	400	160	83	63	23	1.4
14.	Phil	425	275	150	108	25	2.4
15.	JG	398	160	65	55	16	0.5

FMS : Federated Malay States, JG : Java Giant, LO : Lakshadweep Ordinary, CC : Cochin China, NG : New Guinea, Phil : Philippines, ECT : East Coast Tall, VHC-1 : (ECT × MGD), MGD : Malayasian Green Dwarf.

Similarly thirteen hybrids developed from different centres were compared with VHC_1 in underplanted situation. The data in Table 1.20 indicate superiority of ECT × MYD hybrid with annual nut yield of 84 and 9.4 kg copra per palm. (Data in Tables 17 to 20 were drawn from Palm Scientists Meet 1997). (TNAU).

Table 1.20 : Evaluation of Coconut hybrids developed from different Centres at Veppankulam (10½ years old trees) (Underplanting situation)

Sl. No.	Hybrids	Whole nut wt (g)	Dehus-ked nut wt (g)	Kernel wt\nut (g)	Copra wt/nut (g)	Annual yield	
						Nut/ palm	Copra kg/palm
1.	COD × WCT	743	431	225	145	50	4.3
2.	WCT × COD	747	450	230	148	65	6.5
3.	ECT × MYD	765	475	245	157	84	9.4
4.	WCT × GB	725	424	234	142	78	6.2
5.	LO × CC	625	245	186	121	44	2.0
6.	CC × LO	851	495	251	152	36	3.3
7.	LO × Phil	860	502	267	165	47	5.0
8.	ECT × COD	767	483	252	145	74	3.7
9.	GB × ECT	725	361	230	140	39	2.5
10.	GB × LO	605	227	210	132	76	5.4
11.	GB × Phil	625	237	186	120	67	4.9
12.	GB × Fiji	751	452	228	141	53	3.4
13.	CC × Phil	895	525	250	163	44	6.6
14.	VHC_1	765	475	230	145	74	6.4

COD : Chowghat Orange Dwarf, MYD : Malaysian Yellow Dwarf, GB : Gangabondan, WCT : West Coast Tall, ECT : East Coast Tall, LO : Lakshadweep Ordinary, CC : Cochin China, Phil : Philippines, VHC-1 : (ECT × MGD), MGD : Malaysian Green Dwarf.

Performance of VHC_1 Coconut Hybrid in Farmer's Field at Vedaraniam

The Coconut hybrid VHC_1 which was released in 1982 in Tamil Nadu was evaluated in the farmer's field at Vedaraniam (Nagapattinam District). In a holding, 465 plants were planted and the details of yielding trees from the 3rd year after planting (i.e., 1985) upto 7th year (1989) are presented in the Table 1.21.

It may be seen that 77 per cent of the population of VHC-1 have yielded nuts by 7th year after planting, the mean nut yield being 73-year on the 7th year. One tree gave 431 nuts in the 5th year of planting which is a record yield. In this connection it is interesting to note that in VHC_1 hybrid the incidence of Thanjavur wilt disease was nil where banana was riased as intercrop as against 8.6 per cent incidence in pure stand of coconut.

Table 1.21 : Yield performance of VHC_1 hybrid

Year	No. of yielding palms	Per cent to total planted	Total nut yield	Mean no. of nuts per palm per year	Range nuts/ palm/ year	No. of palms yielding more than - nuts/yr/palm			
						100	200	300	400
1985	37	7.9	1842	49.8	2 – 225	6	–	–	–
1986	77	16.4	3021	39.2	2 – 157	6	–	–	–
1987	243	51.7	20985	36.4	2 – 431	87	17	1	1
1988	328	69.8	24500	74.7	1 – 254	113	6	–	–
1989	368	76.6	26141	72.6	3 – 269	107	3	–	–

Source : Project Completion Report.

At Vedaraniam, observations on the performance of VHC-1 Coconut hybrid were recorded for five years in 243 trees out of 465 planted. In the fourth year of planting, 12.7 per cent of the trees commenced bearing. The mean annual yield per palm was found to range from 44.2 nuts in the fifth year to 115.3 nuts in the sixth year. Ten trees have recorded outstanding yield performance. The mean annual yield per palm in these trees were found to range from 132 to 243 nuts. The data have shown the potential and realisable yield in VHC-1 hybrid in the farmer's field under good management. (Ramanathan and Ramachandran, 1993).

Evaluation of Coconut Hybrids at Co-ordinated Centres

Evaluation of hybrids at various centres indicated that Tall × Dwarf hybrids are relatively better than the local Talls in Andhra Pradesh, Maharashtra and Tamil Nadu. Two Tall × Dwarf hybrids were released under the name VHC-1 and VHC-2 from Veppankulam Centre and their performance is given in Table 1.22.

The other promising hybrids at Veppankulam are ECT × GB and MDY × ECT with 115 and 112.5 nuts per palm per year respectively.

Table 1.22 : Yield performance of VHC-1 and VHC-2 at Veppankulam

S.No.	Hybrid/Control	Mean nut yield per palm/ yr. (12 yrs)	Copra content per nut (g)	Oil content (%)	Mean oil yield kg/ha/yr.
1.	VHC-1	130.6	162	68.6	2193
2.	VHC-2	114.0	161	70.2	2551
3.	ECT (Control)	104.8	134	63.5	1566

At Pilicode, Tall × Gangabondan hybrids performed better under rainfed conditions in the laterite soils.

Table 1.23 : Performance of hybrids at Pilicode

S.No.	Hybrid/Control	Annual nut yield per palm	Copra content per nut (g)	Annual copra yield kg/ha	Oil yield kg/ha/yr.
1.	LO × GB (Laksha Ganga)	108.3	194.5	3728	2535
2.	AO × GB (Ananda Ganga)	95.2	216.0	3638	2473
3.	WCT × GB (Kera Ganga)	100.2	201.0	3602	2449
4.	WCT (Control)	64.3	180.0	2059	1400

LO : Lakshadweep Ordinary, AO : Andaman Ordinary, GB : Gangabondan, WCT : West Coast Tall.

Lakshadweep Ordinary × Gangabondan (GB) was recommended for release under the name 'Laksha Ganga' during the Seventh Workshop held at Tiruvanathapuram in 1985.

At Coimbatore, Tall × Dwarf hybrids yielded 551 nuts/palm as cummulative nut yield followed by Dwarf × Tall (463 nuts/ palm) over a period of nine years.

At Ratnagiri, Banaulim × COD hybrid yielded 138 nuts/ palm/year.

Table 1.24 : Performance of T × D hybrid at Ratnagiri

S.No.	Hybrid	Annual nut yield per palm	Copra weight per nut (g)	Oil Content (%)	Annual oil yield kg/ha
1.	Banaulim × COD	138	172	70.8	2946
2.	WCT (Control)	93	182	68.3	2017

COD : Chowghat Orange Dwarf, WCT : West Coast Tall.

At Arsikere, Tall × Dwarf hybrid gave the highest nut yield of 175/palm with 22.0 kg copra/palm/year followed by Lakshadweep Ordinary with 133 nuts and 19 kg copra/palm/year (Rethinam, 1989).

Susceptibility of Coconut Hybrids to Rhinoceros Beetle (*Oryctes rhinoceros*)

With a view to study the fluctuations in the susceptibility of different coconut hybrids, observations on the percentage leaf damage due to beetle attack was recorded from a 13 year old hybrid coconut plantation at the Regional Agricultural Research Station, Pilicode, maintained under rainfed conditions during 1986-88. Among the 16 hybrid combinations, CC × CDO was found to be the most susceptible (21.1%), followed by WCT × LO, AO × CDO and Fiji × CDO. The lowest damage was recorded by CDO × WCT (8%) followed by WCT × CDG and WCT × GB (Sumangala Nambiar, 1988).

Susceptibility to Black Headed Caterpillar

In an experiment conducted to determine the susceptibility and resistance of three coconut varieties, viz., WCT, T×D and DG to *Nephantis serinopa*, it was observed that percentage of both leaf damage and leaflet damage was highest in WCT and lowest in DG. Leaflet characters was a factor that influenced pest damage. The study concluded that DG is relatively tolerant to *N. serinopa* attack and WCT and T × D were susceptible and moderately susceptible, respectively. Hence, DG variety can be exploited as one of the parents for breeding varieties resistant to the pest (Kapadia, 1981).

Hybrid Seed Production Through Seed Gardens

Two general principles are involved in producing high yielding planting material : (1) Prepotency or general combining ability and (2) Hybrid vigour or specific combining ability. A quick method of identifying prepotent palms is based on a high mean number of leaves per plant in the month after transplanting and allows desirable genotypes to be provisionally identified after 40 months. The problem of mass production of seed is also considered. The pollination of selected high yielding palms with pollen from prepotents is suggested as a means of providing good planting material in a short time (Liyanage, D.V., 1972).

In the establishment of isolated seed gardens for hybrid production, the correct balance between Dwarf (seed) parents and Tall (Pollen) parents has to be maintained. Excess Tall (Pollen) palms would be a luxury in view of the space they occupy which could otherwise be planted with Dwarfs. On the other hand, too few or badly distributed (Pollen) palms may lead to a reduced seed set.

Data are presented on three types of distribution patterns : (a) Compact block of Dwarfs surrounded by Tall palms; (b) alternate rows of Dwarfs and Talls and (c) Dwarfs and Talls in the ratio of 25 : 3. (Manthriratnae, 1976).

In Sri Lanka, first specially isolated seed garden for the production of typica × typica was established in 1955 with 800 m jungle as isolation barrier. Then in 1965 the garden was adjusted to produce typica × pumila and pumila (Dwarf) × typica (Tall) hybrids as well.

(a) Degree of spatial isolation

An isolation of 350 m is required to check wind pollination and 800 m for insect (*Apis indica*) pollination.

(b) Ratio of seed to pollen parents

In Ivory Coast one pollinator was used for five mother trees with a spacing varying according to variety of the parents : (7.9 × 7.9 m for typica and 6.7 × 6.7 m for pumila) 81 seed parents : 9 pollen parents were planted in Sri Lanka.

(c) Safeguard in emasculation work

Spathes are split open longitudinally when there is no suitable genetic marker for identification of hybrids. Premature emasculation is done 3-6 days before they would naturally open. At this time all the male flowers are in closed position so that even the act of splitting the spathe does not release any pollen through disturbance. The spikelets are cut about 5 cm above the female flowers and the remaining male flowers are manually stripped off. The male flowers, cut spikelet ends and other debries are then destroyed by burying. Button counts which would give a rough estimate of yield are taken at 12 weeks from the date of emasculation.

Percentage Setting of Emasculated Female Flowers

Inflorescences on 471 pumila palms in the field were emasculated in which 42,074 buttons/nuts were developing 12 weeks from the date of emasculation (mean, 89 nuts/palm) with 95% setting.

Prepotency

A study on parent-progeny analysis for nut yield in East Coast Tall was conducted to assess the prepotency (the ability of the parent to transmit the productive character to its progeny). This is essential to identify ideal parent trees for utilisation in the cross-breeding programme. One among the ten mother palms chosen for the study was found to be the best pre-potent palm. (Ramanathan, 1989b).

Nursery observations on growth rate and seedling vigour as measured by girth at collar and leaf production recorded every month from the time of germination in 599 open pollinated progenies of 16 selected palms of the local West Coast Tall variety have shown highly significant differences in growth rate of progenies between families. Rate of growth increased from fifth month in the progenies of certain families. Correlation of these growth characters from the first to nineth month with those of the tenth month indicated high and positive correlation from fifth month onwards thereby showing that it might be possible to identify palms of superior genetic value from over fifth month for prepotency. These studies also indicated that in areas where young seedlings are preferred for planting, when selection of vigorous growing ones becomes difficult, it is preferable to select seedlings at five month's age based on growth characters like girth at collar and leaf production (Satyabalan and Mathew, 1976).

An investigation was carried out with a view to study the relationship of seednut weight with three seedling characters, namely, girth at collar, height of seedlings and number of leaves in three hybrids, East Coast Tall × Ayiramkachi (ECT × AY), East Coast Tall × Malaysian Dwarf Green (ECT × MDG) and East Coast Tall × Malaysian Dwarf Yellow (ECT × MDY) and in ECT in one year old seedlings. In ECT × AY hybrid, seednut weight was found to be positively associated with seedling height. Girth at collar and number of leaves were not related to seednut weight in this hybrid. While there was significant positive relationship between seednut weight on the one hand and the three seedling characters in ECT × MDG, such relationship was not observed either in ECT × MDY and ECT thereby indicating genotype specificity in such character association (Ramanathan, 1987).

Analysis of variance of data for seedling characters like girth at collar and number of leaves of progeny showed overall means for individual trees as also between hybrids and controls within trees. In almost all cases, the hybrids maintained their superiority in seedling characters over their respective controls though significant differences could be detected only in a few cases (Ninan and Pankajakshan, 1961).

Heritability

Heritability values based on parent progeny regression were estimated for ten characters affecting yield in a population of West Coast Tall. Heritability estimates were low for number of bunches emerged, total number of spikes and spikes with one female flower and high for the remaining characters. It is concluded that selection for large number of spikes with one or two female flowers will contribute toward reducing the instability of yield in this variety (Nambiar *et.al.*, 1970).

Heritability in net assimilation rate (NAR), total chlorophyll content and shoot dry weight have been estimated. The results indicated high heritability for total shoot dry weight, NAR and total chlorophyll content which suggest that in WCT palms the efficiency of dry matter production is a heritable factor (Ramadasan *et.al.*, 1985).

Selection of seed parents on the basis of yield of copra per palm per year, weight per husked nut and setting of female flowers will be highly effective. Selection for number of nuts per palm per year, number of bunches per palm per year, number of female flowers per bunch and number of nuts per bunch may be effective, while selection for early flowering will not be effective (Liyanage *et.al.*, 1988).

Xinia and Combining Ability

Results of preliminary investigations on the effect of different varieties (as pollen parents) on copra content in two trees belonging to Lakshadweep Small and Lakshadweep Micro varieties were reported. In both trees, highly significant differences in copra content of nuts obtained by pollination with different varieties were observed. Variety Kappadam as male showed consistantly better performance in both trees compared to nuts obtained by pollination with Spicata in the first trial and West Coast Tall in the second trial, (which gave the lowest values). Kappadam as male gave an increase of 23.5 g and 22.6 g copra respectively. The percentage increase over controls, of copra content per nut, with Kappadam as male was 8.2% and 35.4% in the first and second trials respectively. It is probable that the increase in copra content resulting from certain combinations is a manifestation of hybrid vigour. There are also indications that varieties with large nut size and high copra content as 'males' give generally better copra content in the resulting nuts. The finding opens up interesting possibilities of improving crop yields by interplanting such breeds as would give maximum copra content per nut on inter-breeding if such breeds are economic themselves and also happen to be high containers the double advantage of increasing crop yields and hybrid seed production could be derived by their interplanting (Ninan *et.al.*, 1963).

The influence of pollen parent on endosperm characters was studied. Forty five Cameroon Red Dwarf mother trees were divided into three groups and fertilized with pollens of (1) Cameroon Red Dwarf (Control); (2) West African Tall (Copra/nut : 200-250g) and (3) Tahiti Tall (Copra/nut : 311 g). The copra/nut obtained from the three crosses were : (1) Control, 220 g, (2) 258 g and (3) 260 g.

These results have shown a very significant effect of the pollen parent on copra/nut. The results of the second cross suggest a possible effect of heterosis.

The relationship $R = \frac{\text{endosperm} \times 100}{(\text{husked fruit-water})}$ was very significantly above the control in the latter. It is probably the same as for copra oil content. (Rognon and DeNuce De Lamothe, 1976).

GENETIC STUDIES

The chromosome number in coconut in 2n : 32. Karyomorphological studies have shown slight differences between Tall and Dwarf palms. Meosis is generally regular in Tall cultivars and in hybrid between Tall and Dwarf, while several irregularities have been reported in the Dwarfs and semi-tall varieties. Pollen fertility is very high in Talls and hybrids as compared to that of Dwarfs and the semi-talls. Dwarfs and Spicata may owe their origin from the Tall as a consequence of inbreeding resulting from self-pollination. Inbreeding depression has been noted in most of the selfed progenies of Tall cultivars. Heterosis is reported in almost all the cross breeding trials.

The genetic components of factors affecting yield has been estimated to be 49% by one method and 62% by another. The high phenotypic value of seed parent may be due to the additive or non-additive effects of genes. Since considerable genetic variation is available for selection in West Coast Tall, selection of trees having characters of low heritability superimposed with high set can help in bringing about yield improvement by mass selection. The selection of mother palms with high yield in hybridization work or for progeny testing appears to be reliable. The selection of genotype with low variance for distribution of female flowers and a large number of spikes each with single female flower may also help to improve and stabilise yield. (Thankamma Pillai *et.al.*, 1976).

Analysis of crosses between six yield groups involving 540 palms drawn from 108 parent palms revealed the existence of substantial additive genetic variation for characters influencing yield. The general combining ability (GCA) of palms yielding over

100 nuts/year was high for number of bunches, female flowers and yield but limited for percentage flower set. Progeny from crosses with pollen from elite palm were superior in all the characters studied to those from open-pollinated or imbred crosses. Pollen from elite palm can be used to upgrade lower yielding groups (Nambiar and Nambiar, 1970).

Genetic Divergence

Genetic divergence using Mahalanobis D^2 technique in 31 cultivars of exotic and indigenous origin was studied for 18 growth and reproductive characters. Cultivars were found to differ significantly for the characters and could be grouped into 11 clusters. Major contributions to divergence were through the nut characters such as weight of kernel and thickness of the shell, kernel and husk. Among the growth characters, girth at collar of three year old seedling, its height and length of leaf, number of leaflets in adult palm and preflowering period were important. Since cultivars from geographically different centres were found in the same cluster it is assumed that geographic origin of a cultivar need not necessarily relate to genetic divergence. The pattern of clustering indicate that there is considerable genetic diversity in the material. Atleast three cultivars formed single number clusters, due to extreme divergence. There is no relationship between the genetic and geographic diversity in coconut (Louis and Chopra, 1990).

Genetic Correlation

Observations were recorded on eight characters and correlated with yield. Data have shown that number of leaflets, total number of leaves, girth of stem at base, length of leaf and girth of stem at crown were found to be positively and significantly correlated with yield in the order mentioned. High degree of positive association was observed between height of stem on the one hand and total number of leaves and girth of stem at base on the other. Stem height was also found to be positively related to number of functional leaves, length of leaf, number of leaflets and breadth of leaf in the order mentioned (Ramanathan, 1984(a)).

Path co-efficient analysis was carried out in population of East Coast Tall cultivar of coconut to identify characters which have direct and indirect influence on yield. Among the twelve characters studied, positive direct effects of number of functional leaves, number of bunches with nuts and number of leaflets on yield indicate the importance of these characters in selection programme (Ramanathan, 1984(b)).

Discriminant function was adopted to formulate selection index used in identifying two economic characters of coconut, namely number of nuts, and copra weight per nut against 17 common growth and reproductive characters. Three characters have common and positive role and two had common but negative role in the selection for either the number of nuts or copra weight per nut. The number of leaves and height of the Juvenile palm were treated as additional requisites for identifying palms with increased number of nuts while the thickness of kernel and shell served as additional characters for identifying palms with good copra weight. The expected genetic advances in selecting palms for the number of nuts and copra weight per nut through the selection index are 36.5% and 68.9% respectively. Index score was the highest for ecotypes from the Philippines (San Ramon) and South America (San Blas) in both selections (Louis and Chopra, 1989).

Selection Index

Heritabilities of, and genetic correlations between various productive traits of the coconut palm have been worked out. They are of considerable importance in a selection programme. Since the yield of nuts and copra are highly correlated positively ($r_0 = 0.79$), and the flowering period and yield of copra are highly correlated negatively ($r_g = -0.81$), selection of seed parents on early flowering and number of nuts will tend to increase the yield of the progeny with respect to copra production. As a number of characters have to be considered in the selection of seed palms, it would be generally useful to construct selection indices. An index 1 using three characters has been computed as follows (Liyanage *et.al.*, 1988).

$$1 = x_1 - 14.70\ x_2 - 4.47\ x_3$$

where x_1 = number of nuts per palm/year

x_2 = weight per husked nut (1b)

x_3 = flowering period of palm (months)

Phyllotaxy

Inheritance studies on two distinct types of phyllotaxy in coconut palm indicated that the F1s segregated for left and right whorls in 1 : 1 ratio, irrespective of left × left, right × left, left × right and right × right combinations. Dominance of either character is ruled out. Asymmetry in coconut palm is decided entirely by probability and does not appear to be controlled by genotypic difference nor does there seems to be cytoplasmic effect. (Louis and Chidambaram, 1976).

Tender Nut Water

The non-reducing sugar content in tender nut water varied considerably (0.04 to 1.68%) in the varieties but the variation for total sugar content was only nominal. The volume of tender nut water showed almost four fold variation. A desired volume of water combined with high sugar content was observed in the varieties, Semi-Tall Red and Semi-Tall Green. The results have shown the possibility for selection of this character (Louis *et.al.*, 1977).

Seedling Vigour

Seednuts of the c.v. West Coast Tall were arranged in 10 groups according to size, shape, and weight. Eleven nuts from each group were planted in the nursery and the following parameters were determined : germination percentage, mean seedling girth at collar, number of leaves, seedling height, percentage of seedlings with six leaves, and width of 6th leaf in 9 and 12 month old seedlings. Irrespective of size and shape, nuts of low or medium weight had the highest germination (99-100%). The mean collar girth was greatest (10.7 cm) in seedlings from egg shaped nuts weighing 960g (group V), whereas seedling height at 9 months was greatest (92 cm) in seedlings from normal round uniform nuts weighing 1150.2 g (group VI). In 12 months old seedlings the height was greatest (147 cm) in group VIII (medium uniform round nuts weighing 345.5 g). The highest percentage of 9 month old seedlings with 6 leaves occurred in group I (small nuts with tapering ends weighing 654.2 g) and group VI. Best results were achieved in normal, round uniform nuts (group VI). (Louis and Annappan, 1985).

Hybrid Vigour was estimated in Juvenile coconut palms through two separate experiments. Importance of exotic and indigenous coconut as parents in producing hybrids to achieve early germination and increased number of leaves per unit height was brought by the experiment. Intermediate values of seedling height with increased leaves in the hybrid indicate the potentiality for more leaves per unit height. Dwarf as female had significant effect on early germination and additive gene effects (GCA) for seedling girth, height and breadth of its laminate leaf. Additive and dominant gene action was true for increased number of leaves in the seedling. (Louis and Chopra, 1988).

Barren Nuts

Studies conducted at Central Plantation Crops Research Institute (CPCRI), Kasaragod have shown that the production of barren nuts formed an erractic feature in coconut. About 69-72% of trees in a plantation produce barren nuts below 3%. Trees producing barren nuts to a larger extent are relatively few. About 2% of the trees are observed to be completely free from signs of barrenness. Six per cent of the trees in a plantation showed this tendency every year. There seems to be some association between the types of barren nuts and the period of production. A predominant tendency is noticed among those trees producing barren nuts during March to June months to produce barren nuts with shell cracked at the apex. Similarly, trees producing more barren nuts during November to February tend to produce barren nuts with shell cracked at the base. Prefertization disturbance resulting in failure of fertilization appears to be responsible for the production of barren nuts with rudimentary solid shell. Defective fertilization (of the flower), post fertilization disturbance and nutritional deficiency are other causative factors. (Mathew, 1976).

A common feature in all coconut tracts is the production of barren nuts which may be partial to complete and estimated to be about 3% of the yield. The occurrence of sterile palms is a rare phenomenon. One palm in Nicobar Island was found to produce sterile male and female flowers with no button shedding. The pigmy sized barren fruit attained a maximum size of 9 cm × 5 cm and a volume of 115 cc after 12 months of growth. All the fruits possessed two internal super-numerary whorls of three alternating carpels each and embryosac was underdeveloped. The male flowers also had three whorls of alternating super-numerary boat-like fleshy sepals, the anthers dehisced normally but were mostly devoid of pollen. (Nair, 1976).

Biennial Bearing

The incidence of the biennial rhythm in coconut yields has been examined on a quantitative basis. Two alternative methods for detecting bienniality in perennial bearers such as coconut are discussed in addition to the methods currently used for orchard crops.

Bienniality is shown to be a significant feature in coconut yields. However, the intensity of bieenial fluctuations is found to be low and therefore of no immediate concern to commercial circles although its importance to research workers on coconut is not to be under estimated. (Abeyavardena, 1962).

Large plots will largely eliminate the biennial effect since 'on' and 'off' phase palms may be assumed to be evenly and randomly distributed in the field. Yield data from 300 palms in a uniformly treated field were used to calculate the index of bienniality for plots ranging in size between 1 and 16 palms. The index was found to decrease with increasing plot sizes as expected, but the decrease was no longer appreciable when the plot size was exceeded 8 to 10 palms while even at a plot size of 16 palms bienniality was not completely eliminated. A minimum plot size of 6 palms is recommended. (Abeywardena, 1965).

An annual fluctuation in yield of coconut palms was classified as low yielders (less than 40 nuts/tree/year), medium yielders (40 to 80 nuts/tree/year) and high yielders (over 80 nuts/

tree/year). The palms in each group were further divided into highly regular bearers (Co-efficient of variation not exceeding 20%) moderately regular bearers (C.V. 20 to 30%) and irregular bearing was high among the low yielders and low among the high yielders. There was a highly significant negative correlation between the mean yield and the C.V. (Satyabalan *et.al.*, 1968).

5. Mass Selection and Strain Building

In perenninal crops like coconut which are propagated through "maternal line selection" according to which attention is paid only to the female parent formed standard practice from very early times. Jack (1930) also pointed out that in the absence of prolonged scientific breeding on coconut, the methodof selection, if carefully performed, should produce good results, but the choice of the parent tree was largely speculative for, although the character of the female parent could be ascertained yet because of its unknown male parent, variation in the offspring was likely to occur to some degree.

Some modifications of this system of selection have been suggested as a means of improvement of heterozygous perennial tree crops like coconut (Liyanage, 1955). This method has been suggested with the assumption that seedligns obtained in this way would be of high yeilding capacity. But Harland (1957) disproved this concept. As the coconut is cross fertilized and consequently the progeny of any given mother is the result of crossing with unknown number of fathers of varying genetic make up, the chances of producing from any given mother a high yielding progeny are thus vey very low and if the progenies are mixed it is likely that they would allow little, if any improvement over a mixed lot of seedlings from uselected or even low yielding mothers.

It is possible to categorise three mass selection variants in the increasing order of effectiveness : (a) selection through selfing, (b) through open pollination and (c) through intercrossing.

The first two methods are of limited interest : Selfing induces a drop in yields without significantly increasing production homogeneity. Open pollination gives very variable results, no doubt linked to lack of pollen origin control. In most favourable cases, the drastic selection necessary for obtaining improvement considerably reduces seed production potential. The only advantage offered by this method is that it is easy to adopt.

Mass selection through intercrossing appears to be much more effective, since it enables stricts choice of male palm, while retaining high seed production potential. No experimental results make it possible to assess the genetic progress to be made. Methods based on the exploitation of heterosis between ecotypes guarantee faster genetic progress.

The various mass selection variants presented where final target was seed production. It is conceivable to use them for improving two populations separately with a view to the subsequent creation of hybrids between improved population.

Varieties Developed through Mass Selection

Systematic multilocational evaluation of indigenous cultivars culminated in the release of Lakshadweep Ordinary (Chandrakalpa) and Pratap (Banawali Green Round) which gave 21.3 and 88.3 per cent higher yields respectively over the local cultivars. Chowghat Orange Dwarf was identified as the best cultivar for use as tender nut (Nair *et.al.*, 1996). VPM-3 was released in 1994 as an outcome of research at Veppankulam (Tamil Nadu) following a selection from Andamand Ordinary (Tall Variety). VPM-3 is tolerant to drought and performs better than ECT under subnormal rainfall.

Results were presented from a progeny trial where mass selection based on weight of husked nuts gave a genetic gain in yield in the progeny. More palms of high breeding value could be obtained by selecting from the best 10% of the palms (on the basis of copra yield), than by taking palms at random. The number of leaves produced in the first 40 months was positively correlated with copra yield at the age of 13 to 14 year. This correlation gave a quick method of detecting palms of high breeding value. (Liyanage, 1964).

6. Close and Linebreeding

The expression of 'close breeding' was derived from the fact that, as far as possible, the pollen produced by individuals in one population was meant to fertilize the flowers of other palms belonging to the same population all of which were derived from nuts harvested from the same individual mother tree (Dwyer, 1938). It was also indicated that in practice it would be difficult to prevent the inter-mixing of the pollen of the several members of the population but some precautions are indicated to eliminate this possibility as far as possible.

Dwyer considered that a continuous systemof close breeding would permit the elimination of inferior strains after each generation until finally only the best strains as determined by performance of yield or any other character under observation are perpetuated. Such a process of elemination, according to him, really amounted to 'strain building' and did not preclude the building up of other improved seed plots from the best strains at any time, without loss of vigour in the resultant progeny. This method offered the quickest means of improvement which can be increased with each generation of further selection.

7. Plant-to-Row Method

In Plant-to-Row method, Dwyer has suggested the selection of seednutsof progeny derived from palms which appeared to be superior by referring previous yield records. Seednuts are sown in nurseries with progeny of the single palms in separate lines for comparison in respect of growth and vigour of seedlings. This sowing in rows makes it possible to eliminate rows of seedling whch are inferrior in performance. After rejecting such rows, planting materials from other row. viz., vigorous seedlings satisfying all the pre-requisites are selected. These seedlings are planted side by side under optimum conditions, so that the heredityof progeny of idividual parenets could be judged to some extent from the very early stage of growth onwards.

8. Clonal Propagation

The coconut palm poses unique problems to the geneticist. The tall, unbrached, slippery and swaying stem deters the scientist from climbing the trunk to carry out breeding work. As the coconut especially the tall variety, is heterozygous, attainment of pure homozygous individual can be accelerated either by increasing the viability of the pollen grains or by propagation of suckering on branching palms whose off-shoots may have synchronous blooming phases. Some methods for effecting vegetative propagation are discussed. The most probable one seems to be the reversal of the spadices of flowers of desirable palms into bulbil shoots, and the bulbils layered and propagated as adults (Davis, 1969(a)).

The only promising method of vegetative propagation of coconut palms is to divide the growting points of young coconut seedling and induce them to produce branches. However, the productive quality of the new seedlings is unknown. It is therefore important to find methods to propagate selected mature palms. It has been observed that floral branches or individual flowers are able to reverse into vegetative shoots (bulbils). (Davis, 1969 (b)).

An abnormal coconut palm in 30 years old plantation at Koilur, near Muthupet in Nagapattinam District was obseved. In the abnormal coconut palm,the spadices, spikes, female flowers and even the male flowers were found to be transformed into vegetative shoot described as 'bulbil shoots'. This is an instance of spontaneous occurrence of mutant form. The bulbil shoots which are developing from the axils of leaves resembled young spadices. But as they enlarged, their flower bearing spikes got suppressed, instead, the development of the subtending bracts got stimulated into large spathes which progressively transformed into pinnate leaves, the shoots ultimately developing into a 'seedling'. Attempts were made to induce roots in the bulbils by separating them from the mother palm and planting them in pots with garden soil and compost mixture at Veppankulam Coconut Research Station, which proved futile (Ramanathan, 1982).

Root-induction in Bulbils

Instances of coconut spadices developing into simple bulbils were reported by many investigators, the important ones being Ridley (1907), Burkill (1915), Furtado and Barreto (1920), Iyengar (1922), Sands (1927), Davis (1967 and 1968), Thomas (1961) and Ramanathan (1982). The older bulbil bore abut 20 to 30 leaves (modified spathes and bracts) most of which were dry. The mature bulbils are curved and the thickness of their 'stem' steadily increases as they grow away from the main stem. The attachment of the bulbil with the main stem resembles that of any normal spadix. Because of the weight of the mini-crown of the bulbils and crowdedness of the main crown, the older bulbils grow slantingly downwards with their tip pointing upwards.

Sudastrip *et.al.,* (1978) at the Industrial Crops Research Institute, Manado, Indonesia ahve observed two coconut palms regularly producing large 'bulbil-shoots' instead of spadices and deviced methods to produce roots from the bulbil-shoots as described here. Bamboo scaffolding and wooden paltform with ladders were built on the palm in November, 1977. Older bulbils were cleared for layering. As the initial step, the curved and bulged portion fo the bulbil was wound closely with paddy straw twisted into ropes. This straw insulation was provided to retain moisture and to enable the rooting zone to have a temperature slightly higher than that outside. A mixture of pulverized volcanic ash, compost and fertilizer mixture (urea, triple superphosphate, muriate of Potash and Kieserite) moistened with water was daubed around the bulbil and wrapped with a perforated plastic sheet. The plastic jacket was covered with jute gunny. Finally, with strips of Jute gunny, the bulbil was firmly tied and further reinforced by coconut coir ropes. Since, the weight of the bulbil after the completion of the operation has increased considerably, it was supported by a bamboo resting on the wooden platform. Each bulbil was numbered. Provision was made for mositening the bulbils daily. December to February is the peak monsoon season for Manado, so artificial watering of the bulbils was discontinued once the heavy monsoon set in. After 147 days from the commencement of layering two mature and adequately rooted bulbils (out of 8 induced for rooting) were detached from mother palm and planted on the ground.

9. Biotechnology Research

To increase the efficiency of genetic diversity, assessment and further safeguard germplasm health, collaboration on biotechnology research between COGENT member countries and advanced laboratories and institutes is being promoted. Collaborative research has been established with Long Ashton

Research Station, UK, to develop applications of RFLP and microsatellite techniques for locating coconut genetic diversity and with Max Planck Institute, Germany on the use of RFLP/PCR techniques for the same purpose. At present genetic diversity assessment is done largely based on morpho-agronomic measurements, a method which is tedious, time-consuming and possibly less precise.

Likewise, research collaboration has also been established with the University of Adelaide, Australia on the use of molecular methods to characterise Coconut Viroids and related molecules and with the Phillippine Coconut Authority to study pathogenicity of CCCvd-like sequences in coconut. It is hoped that the results of these studies would increase the efficiency of disease identification and indexing to promote safe germplasm movement and ensure the health of the conserved germplasm (Batugal, 1996).

Embryo Culture at CPCRI, Kasaragod

The *in Vitro* culture of zygotic embryos solves the problems encountered in transporting and storing coconut seeds especially when obtained under field collecting conditions. Embryo sampling and short term storage methods have been developed. Direct inoculation into culture on media, enabling embryo growth under axenic conditions, has been successfully carried out under field collecting conditions. An alternative method holds embryos in endosperm cores in a salt solution for brief periods before excision and inoculation into culture in a laboratory. The techniques and equipment used are described as well as the subsequent development of the embryos and the transfer of plants to natural growing conditions (Assy Bah *et.al.*, 1989).

Recently the CPCRI, Kasaragod has been able to develop a protocol for field collection of embryos and their storage in sterile water for three months, successful germination and transplantation of plantlets. This is an ideal way for collection of coconut germplasm from distant places at minimum cost (Nair *et. al.*, 1996).

The sexual embryos from the mature and immature nuts were grown on nutrient media to develop into plants. The basic medium composition was not critical for the embryo development. MS, B5 as well as Y3 media supplemented with NAA and BAP were eqully efficient for the germination of embryos. In media with cytokinins, BAP and or kinetin, the shoot development was remarkably fast while the rooting was almost inhibited at levels above 0.5 mg 1^{-1}. In media devoid of cytokinins but with th auxin NAA, the shoot development was slow an haustorium enlarged in to a larg globular structure. As the high level of NAA 5 mg 1^{-1} BAP, there was a callus formation at the haustorium with occasional rhizongensis and abnormal shoot developement. By proper adjustment of the hormone balance coconut plants were produced from excised embryos. These palnts were transferred to a potting mixture of sand and vermiculite moistened with Hoagland's solution. Periodic application of nutrients to the potted plants was necessary for good establishment. More than 60% of the plants were established in the potting mixture (Bavappa, 1986).

Tissue Culture at CPCRI, Kasaragod

Experiments have been carried out to obtain vegetative propagation of coconut palm from explants of variety of tissues including apical meristem tissue, leaf, inflorescence and root tissue. A mineral salt medium has been defined specifically suitable for coconut and perhaps other palm tissue and the organics constituents have been optimised for maximum callus growth. Experiments with callus tissue and organised inflorescence explants are described. Brief mention is also made of experiments to obtain branching in intact palms. (Schewabe, 1976).

Coconut tissue culture work at CPCRI, Kasaragod enabled production of coconut plantlest for the first time from tender leaf segment of West Coast Tall and TxD hybrid seedlings. For producing clonal plantlets, the lower most 10 cm segment of tender leaves was used as the source material. This was surface sterilized by flaming after dipping in 70 per cent alcohol for 2-3 seconds, cut into 5 cm long explants and placed on sterile culture medium containing various nutrients and growth hormones. Activated charcoal was incorporated into the medium to avoid chemical killing of the explants. At the end of 16 weeks incubation, embryoids were initiated in the vascular tissue towards the morphologically lower end of the explant. This initiation time could be be brought down to 3 weeks by adjusting mineral and hormone levels in the medium. Upto 48 embryoids were produced from a single explant. The embryoids wre tripolar with shoot, root and haustorium. Histological studies showed their origin to be directly from the vascular tissues. There was no callus formation at any stage of somatic embryogenesis. The embryoids go separated from the parent leaf segment, after 6-8 weeks and could be made to 'sprout' and develop into plantlets. From the time of collecting the tissue, plantlets of 6-8 cm height with roots could be produced in 6-8 months. Planlets established in the soil are growing well. The response of mature coconut palm tissue was erratic. The responding region in mature trees was found to be limited to 15 to 30 cm segment of the 3rd unopened leaf.

Callus was induced from leaf primodia, subapical zone of stem, veins of leaves, young floral primordia and endosperm of tender nuts. The calli from stem and leaf primordia and leaf veins differentiated roots. The calli or floral primardia did not stand subculturing while endosperm calli could be mutiplied and maintained.

In the attempts of transform the floral primordia into vegetative bulbils, shoot like structure with an indeterminate development pattern producing several leafy (scaly) structure were induced from undifferentiated floral primoridia in the rachilla explants excised out of the inflorescence in the axil of the third leaf from the spindle. These shoot like structures turned green uder light but rooting was erratic. (Bavappa, 1986). Thus, the attempt to develop a protocol for somatic embryogenesis in coconut has not been successful. (Nair *et.al.*, 1996).

Cloning Macapuno

The characterstic of macapuno nut is that the entire cavity is filled with a soft and tender Jelly like endosperm which is a delicacy. In the regular macapuno yielding trees one or two nuts of this type may be found in a bunch of five to seven nuts while the remaining are ordinary nuts. Coconut embryo and tissue culture researches in the Philippines started with the desire to mass propagate the macapuno coconut, an economically important coconut mutant that does not geminate *in sita.* One of the promising technologies Philippines Coconut Research and Development Foundation (PCRDF) is working at the production of the mutant coconut macapuno seedlings through embryo culture. In July 1994, PCRDF began its inhouse macapuno project using the facilities of the PCRDF laboratory. This undertaking hopes to produce seedlings that would mass-propagate coconut trees bearing 60 to 100 per cent macapuno nuts, and realise full commercialisation of the technology.

A primary concern of PCRDF's work which is alongside the efforts of the other agencies like the Philippine Coconut Authority - Albay Research Centre (PCA-ARC) and the VP Los Banos, is to increase the survival rate of the seedlings produced. PCRDF is also show-casing the technology at an experimental farm in Sorsogon in the coconut producing Bicol area (Eleazar Jr., 1996).

Tissue Culture in the United Kingdom

Vegetative propagation of coconut has been achieved through somatic embryogenesis system but the yield of plantlets is low. The present aim is to understand the basic physiological requirements of coconut tissues in order to obtain more successful regeneration of plants.

Inflorescence explants are cultured to give a calloid which produces somatic embryos, but the majority are not normal. A few clonal plants have been produced over several years from more than twenty mother palms, but it is not yet possible to produce large numbers. The somatic embryos often produce haustorial type tissues and calloid can be generated from this using a high auxim treatment.

Wye college has been involved for two decades in the tissue culture of coconut and the research is low funded at the college as part of an aid programme for the Philippines by the German Agency, GTZ. The British Government's Overseas Development Administration has recently given additional funding to the project and European Commission funds are supporting a link with Mexico. (Jennet Blake, 1991).

Embryo-culture in Sri Lanka

Embryo-culture technology has been developed successfully, for use in coconut germ plasm collection and exchange programmes in Sri Lanka. Two packages, one for the laboratory and the other for field collection are offered.

Feasibility of employing embryoculture technology in screening stress-tolerant coconut germ plasm has also been investigaed. The findings were tested on cultivars having different water-stress responses. Under water-deficit conditions, a relatively higher survival rate was observed from putative drought-tolerant cultivars and heavy losses from drought susceptible cultivars, even under moderate stress conditions (Karunaratne, 1991).

References

Anonymous (1985) : Annual Report, 1985 – Central Plantation Crops Research Institute (CPCRI) – Kasaragod.

Anonymous (1993) : Reports of Second Palm Scientists meet held on 3rd May, 1993 at TRRI, Aduturai.

Abeywardena, V. (1962) : Studies on biennial bearing tendency in coconut : 1. The measurement of bienniality in coconut. *Ceylon Coconut Q.* ***13 (3, 4)*** : 112-24.

Abeywardena, V. (1965) : Studies on biennial bearing tendency in Coconut : 2. A minimum plot size for coconut. *Ceylon Coconut Q.* **15 (3, 4)** : 109 - 14.

Aiyadurai, S.G. Mohammad Ibrahim, P.A. and Ramanathan, T. (1959) : Multiple seedling from a germinating coconut (*Cocos nucifera Linn*), *Sci* and *Cult.* **24** : 239-40.

Assy Bah, B., Durand-Gasselin, T. and and Pannetier, C. (1989) : Use of Zygotic embryo culture to collect germ plasm of coconut (*Cocos nucifera* L.). *FAD/IBPGR Pl. Genet Resources News L.* **71** : 4-10.

Balakrishnan, P.C. Devadas, V.S., Unnithan, V.K.G. (1988) : Phenotypic stability of coconut (*Cocos nucifera* L) cultivars for annual yield of nuts. In : National symposium on Coconut Breeding and Management, KAU, Vellanikkara, Trichur, Nov. 23-26, 1988.

Balakrishnan, P.C., Kannan, K. (1988) : Evaluation of certain coconut (*Cocos nucifera* L.) hybrids under rainfed conditions. In : National symposium on Coconut Breeding and Management, KAU, Vellanikkara, Trichur, Nov. 23-26, 1988. Abstracts of papers. p.10.

Batugal, Pons. (1996) : International Coconut Genetic Resources Network (COGENT) Research support to coconut Industry. In : Coconut for Prosperity. (Ed). P.K. Tampan, 249-57.

Bavappa, K.V.A. (1986) : Research at CPCRI. PP. 54-56 Central Plantation Crops Research Institute, Kasaragod.

Bavappa, K.V.A., Nair, M.K. (1989) : Variental improvement in plantation crops. *Journal of Plantation Crops.* **16** : (Suppl) : 349-50.

Bhaskaran, R., Ramanthan, T. (1983) : Variability of coconut varieties and hybrids to natural infection grey blight disease. *Madras agric. J.* **70** (10) : 685-86.

Burkill, I.H, (1915) : An abnomality in the Coconut Palm. *J. Royal Asvatic Soc. Strait Br.* 69 : 15.

Davis, T.A. (1967) : Foliation of coconut spadices and flowers. *Oleagineux.* **22** : 19-23.

Davis, T.A. (1968) : Difficulties in gentic improvements of the coconut : a promising alternative method. *Indian J. Genet.* and *Pl. Breed.* **28A** : 154-164.

Davis, T.A. (1969a) : Clonal propagation of the coconuts. *J.Agric. Soc., Trinidad Tobago* 69(3) : 810 also in World Crops. 21 : 253-55.

Davis, T.A. (1969b) : Prospects of clonal propagation of the coconut. *Ceylon Coconut Plant Rev* **6(i)** : 1-5.

Dwyer, R.E.P. (1938) : Coconut improvement by seed selection and plant breeding. *New Guinea Agric Gaz.* **4** : 24-102.

Eleazar Jr. Jose R. (1996) : Coconut Research in the Phillippines. In : Coconut for prosperity. (Ed). P.K. Tampan, 215-20.

Fremond, Y., Ziller, R., de Nuce de Lamothe, M, (1966) : Le Cocotier. Maison neuve and Lavoise, Paris.

Furtado, C.X. and Barreto, P.C. (1920) : Abnormal inflorescences. *Poona Agric Col. Mag.* **12** : 81-83.

Gangolly, S.R., Satyabalan, K., Pandalai, K.M. (1957) : Varieties of the coconut. *Indian Coconut J.* **10** : 3-28.

Gill, K.S. (1993) : Sustainability in oil seed production systems. In : National Seminar on Oil seeds. Research and Development in India : Status and strategies. Extended summaries. P.94. August 2-5, 1993, Hyderabad.

Harland, S.C. (1957) : The improvement of the Coconut Research Institute, Ceylon.

Harries, H.C. (1970) : The Malayan Dwarf supersedes the Jamaica Tall Coconut-1. Reputation and performance. *Oleagineux.* **25** (10) : 527-31.

Harries, H.C. (1971) : Coconut varieties in America. *Oleagineux* **26**(4) : 235-42.

Harries, H.C. (1974) : The performance of F1 hybrid Coconut in Jamica. J. *Plant. Crops* (India) **2**(2) : 15.20 Also in *Oleageineux* **31** (7) : 359.

Harries, H.C. (1978) : The evolution, dissemination and classification of *cocos nucifera* L. *Bot. Rev.* **44** : 265-319.

Harries, H.C., Rommey, D.H. (1974) : Maypan; an F_1 hybrid coconut variety for commercial production in Jamaica. *World Crops.* **26** (3) : 110-11.

Iyengar, M.O.P. (1922) : Note on bulbiferous coconut tree for Malabar. *J. Indian Bot. Soc* **3** : 289-91.

Iznaga, F.A. Tabayoyong, F.T. (Eds) (1967) : Varieties of coconut grown in the Phillippines. *ESSO Agroserv. Bull* **9** : 9-11.

Jack, H.W. (1930) : Improvement of the coconut crop by selection. *Malay agric J.* **18** : 30-39.

Jack, H.W. and Sands, W.N. (1929) : Observation on the dwarf coconut palms in Malaya. *Malay agric. J.* **17**(6) : 140-65.

Jennet Blake (1991) : Current status of coconut tissue culture in th United Kingdom. Abstract P.31. In : International symposium on Coconut Research and Development-II. 26-29 November, 1991 held at CPCRI, Kasaragod.

Kapadia, M.N. (1981) : Susceptibilityof three cultivated coconut varieties to *Nephantis serinopa* Meyr. *Indian Cocon. J.* **12** (4) : 4-5.

Karunarathne, S.M. (1991) : Current status of embryo and tissue culture in Coconut in Sri Lanka, Abstr. P.33. In International Symposium on Coconut Research and Development-II, 26-29, November, 1991 held at CPCRI, Kasaragod.

Liyanage, D.V. (1955) : Planting materials for coconut. Coconut Conference, Lunewila.

Liyanage, D.V. (1957) : Annual report of the Coconut Research Institute, Ceylon for 1957.

Liyanage, D.V. (1958) : Varieties and forms of Coconut Palm grown in Ceylon. *Ceylon Coconut Q.* **9** : 1-10.

Liyanage, D.V. (1964) : Mass selection and progeny testing in coconuts, pap presented at the second sess FAO Tech working party on coconut prod prot process, Nov 30 - Dec.8, 1964. P.11.

Liyanage, D.V. (1972) : Production of improved coconut seed by hybridization. *Oleagineux* **27** (12) : 597-99.

Liyanage, M de S. (1996) : Coconut Research in Sri Lanka. in : P.K. Thampan (Ed) Coconut for prosperity PP. 221-37.

Liyanage, D.V. Wickramaratne, M.R.T. Jayasekara., C. (1988) : Coconut breeding in Sri Lanka: A. review. *Cocos* **6** : 1-26.

Louis, I. Henry and Annappan, R.S. (1985) : Observation on the seedling vigour in relation to the size and shpe of seednut in Tall variety of cocontu palm. *Indian Coconut J.* **16**(7) : 5-11.

Louis, I. Henry and Chidambaram. (1976) : Inheritance studies on phyllotaxy of coconut palm. *Ceylon Cocon. Q.* **27** (3-4) : 22-24.

Louis, I. Henry and Chopra, V.L. (1988) : Exploiting heterosis in coconut palms. In : National symposium on Coconut Breeding and Management. KAU, Vellanikkara, Trichur, Nov. 23-26, 1988.

Louis, I. Henry and Chopra, V.L. (1989) : Selection index in coconut palm. *Philipp J.* of *Cocon. Studies* **14**(1) : 10-13.

Louis, I. Henry and Chopra, V.L. (1990). Genetic divergence in coconut (*Cocos nucifera* L). In : Biological Approaches and Evolutionary trends in plant. International symposium of plant Bio-systematics, 4th, Tokyo, 1990. Proceedings of symposium pp. 11-20.

Louis, I. Henry; Vijayan,K.P.; Peter, S.D. and Jayaraman, C. (1977) : A study on the variation of tendernut characters among coconut varieties, J. *Plantn. Crops* **5**(1) : 59-60.

Manthri Rathne, M.A.P. (1972) : The performance of dwarfs (*Cocos nucifera* L.Var. 'nana') as plantation crops in Ceylon. *Ceylon Cccont Q.* **23** (3,4) : 92-99.

Manthri Ratne, M.A.P. (1976) : The mechanics of hybrid seed production throgh seed gardens. In : Abstr. Pap Int Symp Coconut Res. Dev., Dec. 28-31, 1976, Kosaragod.

Mathew, C. 1976. Barren nut production in coconut. *Coconut Bull* 6 (12): 1-2.

Michellia Darwis, (1996) : Coconut Research in Indonesia. In : Coconut for prospertiy (Ed) P.K. Thampan. 207-214.

Nair, M.K. Ratnambal, M.J. and Gopalasundaram (1996) : Coconut Research in India. In : Coconut for prosperity, (Ed) P.K. Thampan. pp. 181-198.

Nair, R.B. (1976) : A coconut palm prodcing pigmy sized barren fruits. In: *Abstr. Pap Int Symp coconut Res Dev.* Dec. 28-31.

Nallathamabi, G., Raveendran. T.S., Ramanthan, T., Vijayaraghavan. H. (1986) : Exploitation of heteroisis for oil content in coconut (*Cocos nucifera* L.). *Indian Coconut J.* **17**(4) : 1-2.

Nambiar, M.C., Mathew, J., Kutty, S.S. (1970) : Inheritance of nut production in coconut. *Indian J. Genet Plant Breed.* **30**(3) : 599-603.

Nambiar, M.C. and Nambiar, K.P.P. (1970) : Genetic analysis of yield attributes in *cocos nucifera* L, Var. West Coast Tall. *Euphytica* **19**:543-51.

Narayana, G.V. and John, C.M. (1949) : Varieties and forms of the coconut. *Madras Agric J.* **36** : 349-366.

Ninan, C.A. and Pankajakshan, A.S. (1961) : Progeny studies in coconuts. 1. Relationship between parent yield and seedling characters of progeny with special reference to open-pollinated and hybrid progenies of West Coast Talls and its bearing on the concept of prepotency in coconuts. *Indian Coconut J* . **14**(3) : 100-107.

Ninan, C.A., Pankajakshan, A.S. and Radhakrishan, V. (1963) : Preliminary observations on the influence of pollen parent on copra content in cocontus. (*Cocos nucifera* L) *Indian Coconut J.* **16**(4) : 174-78.

Patel, J.S. (1937) : Coconut breeding. *Proc. Assoc Econ Biol.* **5** : 1-16.

Pons Batugal. 1996, International Coconut genetic resources network (COGENT) Research support to coconut industry. In : Coconut for prosperity, (Ed) P.K. Tampan. 249-57.

Rajagopal, V., Kasturibai, K.V., Voleti, S.R. and Shivashankar, S. (1988a). Water stress in coconut palms. Paper presented in the International Cong of Plant physiology, New Delhi, Jan. 15-20, 1988.

Rajagopal, V. Kasturibai, K.V. and Voleti, S.R. (1990) : Screening of Coconut genotypes for drought tolerance. *Oleagineux,* **45**(5) : 215-233.

Rajagopal, V., Shivashankar, S., Kasturibai, K.V. and Voleti, S.R. (1988b). Leaf water potential as an index of drought tolerance in Coconut. (*Cocos nucifera* L.) *Pl. Physiol Biochem* **15** : 80-86.

Rajagopal, V., Voleti, S.R., Kasturibai, K.V. and Shivashankar, S. (1988c). Physiological and biochemical criteria for breeding drought tolerance in coconuts. In : National Symposium in Coconut Breeding and management, KAU, Vellanikara, Nov. 23-26, 1988. Abstracts of Papers p. 18.

Ramachandran, T.K., Raveendran, T.S. Ramanathan, T., Louis, I., Henry, (1990) : VHC-2 A new TxD hyrbid coconut for Tamil Nadu. *Madras Agric J.* **77** (5 and 6) : 211-16.

Ramadasan, A., Kasturibai, K.V. Shivashankar, S. and Vijayakumar, K. (1985) : Heritability of seedling vigour in coconut seedling. *J. Plantn. Crops:* **13**(2) : 136-38.

Ramanathan, T. (1982) : Clones in seed propagated coconut. *Kisan Wld.* **9**(5) : 32-33.

Ramanathan, T. (1984a) : Character association in Coconut. *Madras Agric J.* **71**(3) : 191-92.

Ramanathan, T. (1984b) : Path Coefficient analysis for yield in East Coast Tall coconuts. In : *Sixth symp plantn. crops,* Kottayam, Dec. 16-20, 1984.

Ramanathan T. (1987) : Association of seednut weight with seedling characters in coconut. *Indian Coconut J.* **18**(4) : 3-4.

Ramanathan, T. (1989a) : Heterosis breeding in coconut. (*Cocos nucifera* Linn). *Indian Coconut J.* **19**(10) : 5-10.

Ramanathan, T. (1989b) : Parent progeny analysis for nut yield in East Coast Tall. *Indian Coconut J.* 20(4) : 3-4.

Ramanathan, T., and Louis, I. Henry (1983) : Heterosis in Coconut. In: Genetics of heterotic system in crop plants. PP. 140-47. Pre Congress Symposium XV International Genetic Congress held at School of Genetics, TNAU, coimbatore, Dec. 7-9, 1983.

Ramanathan, T. Louis, I. Henry., Navakodi, K. (1984) : High yielding VHC_1 Coconut hybrid. In : Proceed Fifth Annual Symposium on plantation crops held at CPCRI, Kasaragod. 15-18 Dec., 1982 pp. 125-130.

Ramanathan, T., Nallathambi, G. (1988) : Performance of unirrigated East Coast Tall x Dwarf Green and East Coast Tall x Chowgaht Dwarf Orange Coconut hybrids. *India Coconut J.* **18**(12) : 3-4.

Ramanathan T., and Ramachandran, T.K. (1993) : Evaluation of VHC_1 Coconut hybrid in Farmer's field. *Indian Coconut J.* **24**(5) : 10-12.

Ramanathan, T., Thangavelu, S., Sridharan, C.S.; Alarmelu, S. (1992) : Performance of coconut cultivars and hybrids under semidry condition. *Indian Coconut J.* **23** (6) : 9-11.

Rao, E.V.V.B., (1989) : Plantation Crops genetic resource research in India. *J. Plantn Crops.* 16 (Suppl) 303-12.

Rao, E.V.V.B; Pillai, R.V. (1984) : Characterization of coconut germplasm based on fruit component analysis. In : Proceed Fifth annual symp plants crops. CPCRI, Kasaragod, 15-18 Dec., 1982. pp. 112-24.

Rao, E.V.V.B., Pillai, R.V., Mathew, Jacob (1988) : Relative drought tolerance and productivity of released coconut hybrids. In : National Symposium on Coconut Breeding and Management. KAU, Vellanikkara, Nov. 23-26, 1988. Abstr. of Papers. p. 20.

Raveendran, T.S.; Ramanathan, T; Nallathambi, G; Vijayaragavan, H. (1987) : Metroglyph anaylsis in coconut (*Cocos nucifera* L). *Cocos* 5 : 32-38.

Raveendran, T.S., Vijararaghavan, H.; Ramachandran, T.K. (1989) : Some physiologic aspects and production trends of certain coconut hybrids and their parents. *Cocos.* **7** : 36-41.

Reddy, P.S. (1995) : Oilseeds - Scope for big production rise. In : *"The Hindu"*, Survey of Indian Agriculture - 1995, P. 71-79.

Rethinam, P. 1989, Research high lights of All India Co-ordinated Research Project on palms. 1972-1989. Technical Bulletin - 21. CPCRI, Kasaragod.

Rethinam, P. (1993) : Oil bearing tree species. In : Natn Semin Oilseed Res and Dev in India : Status and strategies, Aug. 2-5, 1993. P. 24-27 (Indian Society of Oil Seeds Research, Rajendranagar, Hyderabad).

Ridley, H.N. (1907) : Branching in palms. *Ann Bot Lond.* **21** : 415-422.

Rognon, F. (1996) : Coconut Research by CIRAD-CP. In : Coconut for prosperity (Ed) P.K. Thampan, pp. 157-63.

Rognon, F., De Nuce De Lamothe, M. (1976) : Xinia and combining ability in the coconut palm. In : *Abstr Pap Int Symp Coconut Res. Dev.* Dec. 28-31, 1976, Kasaragod.

Sands, W.N. (1927) : Abnormalities in coconut palm *Malay Agric. J.* **15** 290-293.

Satyabalan, K., Mathew, J. (1976) : Identification of prepotent palms in West Coast Tall variety based on growth of progeny in the nursery. In : Abstr. Paper Int. Symp Coconut Res Dev. Dec. 28-31, 1976 Kasaragod.

Sathyabalan, K., Nambiar, M.C. (1968) : The present position of coconut improvement by hybridization in India. Paper presented at the third sess FAO The working party Coconut prod prot. process, Sept. 9-19, 1968, Jakarta, Indonesia, 5P.

Satyabalan, K., Ratnam, T.C., Kunjab, T.V. (1970) : Hybrid vigour in nut and copra characters of coconut hybrids. *Indian J. Agric. Sci.* **40** (12) : 1088-93.

Satyabalan, K., Sankar, N., Chami, P. (1968) : Studies on bearing tendency of the coconut palm (*Cocos nucifera* L) 1. The character irregular bearing and its incidence in different yield groups. *Trop Agric* **45**(1) : 67-71.

Schwabe, W.W. 1976. Attempts at Vegetative propagation of coconut palm. In : Abstr. Pap Int Symp Coconut Res Dev. Dec. 28-31, 1976, Kasaragod.

Sudasrip, H., Kaat, H; DAVIS. A. (1978) : Clonal propagation of the coconut via bulbils, *Philippine J. Coconut Studies* 3(3) : 5-14.

Sumangala S. Nambiar, (1988) : Susceptibility of hybrid coconut varieties to *Oryctes rhyniceros* Linn (Scarabaeidae) under rainfed conditions at Pilicode. In : National Symposium on coconut breeding and management, KAU, Vellanikkara, Nov. 23-26, 1988.

Thampan, P.K. (1997) : Coconut-Gains from organic farming. *'The Hindu'*, Survey of Indian Agriculture, pp. 89-93.

Thankamma Pillai, P.K.; Vijayakumar, G; Nambiar, M.C. (1976) : Cytogenetic and genetic studies in coconut : A review. In : Abstr Pap Int Symp Coconut Res Dev. Dec. 28-31, 1976, Kasaragod.

Thomas, C.A. (1961) : Coconut bulbils with roots. *Indian Coconut J.* **14**: 132-134.

2

Protection Technologies of Coconut

PESTS AND DISEASES OF COCONUT

The gains achieve through genetic improvement of coconut can be sustained only if the yield reduction due to pests and diseases affecting coconut is prevented. Hence, a detailed review of research on these aspects is considered relevant and essential.

I. Pests

Insect pests affecting coconut include, Red palm weevil, rhinoceros beetle, blackheaded cater pillar, coconut scale, white grub, bag worm, termite, coreid bug. scolytid bark borer, and weevil Diocalandra. Apart from these which cause varying degree of damage, mites, rematodes and rats also have a role to paly in loss of productivity.

(A) Insect Pests

1. Red Palm Weevil (*Rhynchophorus ferrugineus* F.)

The red palm weevil is a dreaded pest of coconut palm, capable of killing the palm outfright. Though different methods of management have been recommended agianst this pest, any single method has not been proved to be effective in combating the pest infestation. Timely adoption of curative chemical control can protect the palms from pest infestation. But being a tissue borer, the infestation is quite often detected at later stages only and as such it would be difficult to save the palm with curative treatment. Hence, an integrated method of pest management system was developed. The schedules included adoption of plant and field sanitation, prophylactic method of filling all the leaf axils of palms with BHC 10% dust + sand, curative chemical treatment, prevention of pest entry through cut ends and wounds and use of attractants. (Abraham *et.al.*, 1989).

Habit : The adult weevils locate and breed on damaged palms. Both adult weevils and their grubs are always found together in the affected palm. Adults move from one palm to another. In heavily infested areas, adult weevils are attracted to open wounds on trunks of palms. Adult weevil is a reddish brown cylindrical ionsect with a long prominant snout. The grub is stout, legless and yellowish with a brown head.

Symptoms and Detection : It is rather difficult to detect the presence of weevil grubs inside the trunks of palms as the affected palms collapse without any external sysmptoms of attack in many cases. However, the presence of holes on the stem with chewed fibrous material protruding out will give a clue. Exudation of redish brown liquid can also be seen from the holes or cracks in severely infested palms, central leaves wither and the palm collapses.

Young palms should be inspected periodically for cracks and wound on the palms to facilitate detection of infestation at the earliest. Unnecessary wounding of young palms by way of harrowing, chopping of stems and other careless operations should be avoided, since all wound are attractive for egg laying. All such wounds should be smeared with tar.

Work done at Kerala Agricultural University, Trivandrum, revealed detection of higly potent polyhedrosis virus (CPV) in stages of the red palm weevil, *Rhyncophorus ferrugineus* F. It infects all life stages of the pest, including the adult. Infection in the late grub stage resulted in malformed adults and suppressed the insect population drastically. The midgut of infected insects was enlarged and filled with thousand of polyhedral inclusion bodies (PIBs) visible under light microscope. Electron microscopic studies revealed surface projection of viral bodies characteristic of CPV (Gopinathan *et.al.*, 1990).

Traps and attractants, by which adults can be trapped and killed have been proved to be effective components of integrated pest management. Log traps consists of tender coconut stems, 50 cm long and split longitudinally were found to be superior to metal traps. The cut surfaces were treated with candidate attractants.

Out of sixteen combinations tested in the field, logs treated with coconut toddy + yeast + acetic acid were the best, followed by coconut toddy + acetic acid, pineapple + molasses + yeast. The use of coconut log traps treated with the above combination is recommended as one of the effective components of integrated pest management. (Kurian *et.al.*, 1984).

Feasibility of using sterile male release method for the control of red palm weevil, was investigated. A total of over 5000 radio sterilised males were released in an 800 acre coconut plantation near Kayangulam in Kerala. Data on the fertility pattern of native females are presented and the factors affecting the efficacy of sterile male release of control of this insect are discussed (Rahalkar *et.al.*, 1977).

Relative effectiveness of root feeding of seven insecticides was assessed. None of the insecticides except, monocrotophos at 10 ml/palm diluted with equal quantity of water was found to give optimum lethal dose. (Rao *et.al.*, 1989).

Other control measures suggested are :

(i) Remove and burn all wilting or damaged palms in coconut gardens to prevent further perpetuation of the pest.

(ii) Avoid injuries on stems of palms as the wounds may serve as oviposition sites for the weevil.

(iii) Treat wound (if inflicted accidentally) with BHC 0.1% solution or tar.

(iv) Avoid cutting of green leaves. If needed they should be cut about 120 cm away from the stem.

(v) Fill the crown and the top most three leaves axil with a mixture of fine sand and BHC 10% (in equal proportion) or chlordane 5D (1:1 by volume) once in three months to prevent the attack of rhinoceros beetle damage in which the red palm weevil lay eggs.

(vi) Plug all holes and inject pyrocone E or carbaryl at 1% or 10 ml of Manocrotophos or 5 ml of Monocrotophos + 5 ml Dichlorvos into the stem by drilling a hole above the points of attack and plug the hole with cement.

2. Rhinoceros Beetle

Rhinoceros beetle *Oryctes rhinoceros* L. attacks coconut palm, oil palm (*Elaeis guineensis*), palmyra palm (*Borassus flabellifer*) and date palm (*Phoenix dactylifera*).

The adult beetles bore through unopened fronds and spathes. The affected frond when fully opened will have a fan like appearance showing characteristic inverted V shape geometric cuts. The other visible symptoms are holes in the axils of leaves, small lumps of chewed up fibrous material at the entrance of the bore holes. In severe cases, the crown is disfigured. Infestation on the spathes destroys the inflorescene and thus prevents nut production.

Adult beetles are dark brown to black with backward directed horn-like projection on the dorsal portion of the head. The horn is more prominent in males. They breed in decaying organic debris, farm yard manure, dead coconut stumps, logs and compost. The females lay upto 140 eggs in her life time. The life cycle is completed within six months.

Research on Control Measures

(1) *Biological* : Release of the exotic predacious reduviid *platymeris laevicollis* Distant at three places in India since 1968 for the biological control of *Orycters rhinoceros* L on coconut are described. Although the predators killed many beetles, infestation tended to increase when releases ceases. Releases were continued at one site in Karnataka (Anthony *et.al.*, 1979).

A baculovirus disease was found in *Oryctes rhinoceros* in Kerala. The symptom confirmed to those reported from South-East Asia and South Pacific Islands. The disease was diagnosed and confirmed by visual symptoms of midgut, smear of midgut contents, immuno-osmophoresis, bioassay in grubs and beetles and electron microscopy of midgut section of the beętle. The smear test and immuno-osmophoresis were found to be ideal diagnostic method for routine screening. The bacilo form baculovirus particles measured 215 - 260 x 77 - 198 mm.

The population of *Oryctes rhinoceros* L. has been substantially reduced in Minicoy Island, Lakshadweep, through the use of a virus disease of the pest. *O. rhinoceros* beetles were infected with *oryctes baculovirus* (30 LD 50 dose/ml. of inoculum) and released into the pest population in Minicoy Island in April, 1983. The virus was previsoul;y isolated from field collected, diseased *O. rhinoceros* beetle from Kayankulam in Allepey district, Kerala. Prerelease observations at Minicoy in April, 1983 revealed high incidence of pest damage to coconut in the absence of the viral disease in the pest population.Post-release observations recorded after two years revealed spread of the virus disease to the subsequent generations of beetles and larvae in breeding sites. The resultant epizootic in the pest population drastically reduced the larval popualtion in breeding sites, leading to a substantial reduction in the damge to coconut palms in experimental plot (Mohan *et.al.,* 1989).

Biochemical : Characteristics of cuticular phenoloxidases of the coconut pest, *Oryctes rhinoceros* were studied biochemically with respect to substrate specificity, pH optima, thermostability and the behaviour of the enzyme in the presence of various chemicals. Biochemicals studies indicate that the larval cuticle contains tyrosinase (EC. 1.10.3.1 ortho diphenol : O_2 oxido reductase) and laccase type of enzyme (FC, 1.10.3.2 paradiphenol; O_2 oxido reductase). These enzymes showed similarity to enzyme type A and C with respect to the enzyme activity in the larval cuticle of *callphora viscina.* In the electrophoretic study, two fractions are discernable in the gel which cofirms the occurrence of two enzyme namely tyrosinase and lacasse types. (Jeyaraj *et.al.,* 1986).

In the scarabaeid beetle *oryctes rhinoceros,* most of the 'A' type neurosecretory cells the brain appear to be elaborate a diuretic principle which is stored in the corpus cardiacum. However, one pair of the 'A' cells among them lying close to the median line elaborate on antidiuretic principle. (Kannan and Prabhu, 1985).

Protease, amylase, lipase and trehalase are present in the larval and adult midguts of *oryctes rhinoceros.* The presence of trehalase in the pupal midgut suggests that food digestion is not the normal function of gut trehalase. Quantitative studies reveals that protease and amylase in the third instar larvae reach very low level as the larvae become older and consume little or no food. These two enzymes are not measurable in the non-feeding prepupa and pupa and reappear in the adult with commencement of feeding. Secretion of digestive enzymes is correlated with feeding (Sree Kumar and Prabhu, 1988(a)).

The effect of midgut ephithelial extract on digestive enzyme secretion in the third instar larva of *Oryctes rhinoceros* was studied. Midguts dissected out from third instar larvae were cut at both ends, emptied their contents, washed in insect saline, ligated at the open ends and incubated in a medium containing extract of 2 midgut epithelila/90 ml incubation solution. As control midgut preparations were incubated in insect saline without midgut epithelial extract. After 30 min of incubation, the content of the midgut preparations incubated with midgut ephithelial extract showed increased secretion of protease and amylase when compared with controls. Digestive enzyme release into gut lumen appears to be due to the action of a hormone present in the midgut epithelium. (Sree Kumar and Prabhu, 1988(b)).

Chemical Repellant : The effectiveness of application of 1:1 mixtures of sand and the granules of four different insecticides at various rates to both sides of the crown region or the first leaf axil of coconut palms for control of *Oryctes rhinoceros* was determined in field tests in Tamil Naud in 1978. Comparrison was made with treatment having 10% dust of HCH at 0.1 Kg/palm. Treatment with 10% granules of phorate gave complete protection upto 60 days on 3-5 year old palms when aplied at 5g/palm and on 1-2 year old seedlings when applied at 0.5g/palm. (Jayaraman, 1985).

Different insecticide formulations were tried and evaluated in the base of the interspaces of leaf sheath of three top most leaves in the crown of 4 to 5 old coconut palms. All the chemical treatments were found distinctly superior to untreated control. Application of naphthalene balls (weighing 3.5 g each @ 3 per palm at 45 days interval protected palms effectively against rhinoceros beetle as that of HCH 10% + sand 1 : 1 ratio. (Sadakatulla and Ramachandran, 1990(b)).

The following pest management measures are suggested :

(i) Remove and burn all dead coconut trees in the garden (which are likely to serve as a good breeding ground) to maintain good sanitation.

(ii) Remove the beetles from the crown by a beetle hook (an iron rod 75 cm long and 5mm thick, with a hook like tip at one end and filling the hole and axils around the spindle with a mixture of fine sand and BHC 10D or Chlordane 5D or Aldrin 5D (1:1 by volume), once in three months, particularly before and after the monsoon months to check the damge by adults. Alternatively application of 10% granules of phorate at 5g/palm at 60 days interval or naphthalene ball at 3 per palm at 45 days interval is recommended.

(iii) Collect and destroy various bio-stages of the beetle from the manure pits (breeding ground of the pest) whenever manure is lifted from the pits.

(iv) Drench the manure pits with BHC 0.1% solution once in three months to kill the grubs.

(v) Set up light traps following the first rains in summer to attract and kill the adult beetles.

(vi) Soak castor oil cake in small mud pots and keep them in the coconut gardens to attract and kill the adults.

(vii) Treat the longitudinally split tender coconut stem or green petiole of fronds with fresh toddy and keep them in the garden to attract and trap the beetles.

(viii) The green muscardine fungus *Metarhizium anisopliae* which infects grubs and causes heavy mortality during the monsoon. This can be used for biological suppression.

(ix) Recently, a virus, *Oryctes baculovirus* has been employed for biological suppression. The pathogen reduces fecundity of infected female beetles.

(x) Release of exotic predator, reduviid *platymeris laevicollis* was found to reduce infestation.

3. Black-Headed Caterpillar

Black-headed caterpillar, *Opisina arenosella* walker (*Nephantis serinopa* Meyr Lepidoptera : Cryptophasidae) is an important defoliating pest of coconut in all coconut growing states. The coastal tracts of coconut growing regions experience heavy infestation. However, in recent years the pest has appeared in enormous proportions in certain interrior parts also, causing severe economic loss to the growers.

The caterpillars feed on the chlorophyl containing parenchymatus tissues of the leaflets. They live in galleries made of silk and reinforced with leaf scrapes and excreta. Infestation starts from lower older leaves. It slowly spreads to upper leaves covering the whole frond. The pest attack can be seen at any stage of the palm. Feeding results in straw coloured spots on the upper surfcace of the leaflets. These spots unite and ultimately the leaf dries. This results in a scorching appearance which can be identified even from a distance. Severe infestation reduces the functional leaf area and may result in drying of fronds adffecting the nut yield considerably.

The caterpillar multiplies prolifically during summer. With the onset of mosoon the pest population declines. Its adult is a straw coloured, medium sized moth often seen resting on the lower surface of the leaves damaged during the day. It flies at night and lays eggs in small gourps on the lower surface of the scraped and damage leaflets. A single female can lay 150-300 eggs during its life span of 6 - 7 days. Incubation period lasts for 5 to 7 days. The larvae feed on the undersurface of leaves, gregariously. Fully fed larva constructs a gallery of silk and frass in which it retreats. The head is black and the body is light green with reddish brown stripes all over. The larval period comprising five instars is completed in 5 to 7 weeks. Pupation takes place in cocoon within the larval gallery. Adult emerges out of brown pupa in 7 to 8 days. Thus one life cycle is completed in 7 to 10 weeks. About six overlapping generations can occur in a year.

Results of Research on Parasites, Predators, Micro-Organisms, Insecticides, and Integrated Management of Black Headed Caterpillar

Predators : A species of *cardiastethus* (Hemiptera : Anthocoridae) predatory on the eggs and early larval instars of *N. serinopa* is recorded in Kerala. The predatory bugs have five nymphal instars all of which occupy an average of 18 to 22 days.

The immature stages as well as the adults are predaceous but are non-specific having a range of hosts. The adults live on an average one month but may survive upto three months. Ovipositions occur through out the life of mature females, each laying about 17-20 eggs. (Abdurahiman *et.al.*, 1982).

The biology of species of *Cardiastethus* near *C. nazarenus*, a predator of the xyloryctid coconut pest *Opisina arenosela* in Gujarat, was studied in the laboratory in 1981. First instar larvae *Corcyra cephalonica* were provided as food. The incubation period was 2 to 5 days with an average of 3.4. The nymphal instars averaged 4.5, 2.5, 2.0, 2.4 and 4.4 days respectively. The total nymphal period averaged 15.6 days. Each female laid 8 to 23 eggs with an average of 14.3. (Kapadia, 1987).

Short description of the life history, reproduction and predatory behaviour of two beetles, viz., *Parena nigrolineata* chaud and *Calleida splendidula* (F), which are predators on the coconut catrpillar, *Nephantis serinopa* Meyr., are given (Mohamed *et.al.*, 1982).

A method to rear *Parena nigrolineata* in the laboratory was developed. Biology and feeding potential of the predator both in the adult and immature stage are discussed. The rate of predation in the grub phase worked out to be 0.32 fourth instar caterpillar of *O. arenosella* per grub/day. Adult female beetle fed 0.34 and male beetle 0.31 fourth instar caterpillar/beetle/day. A scelionid *Xenomerus* sp was recorded as an egg parasitoid of *P. nigrolineata* in the field. Intesity of its natural parasitism was quite high during rainy season and this was responsbile for minimising the population of the beetle in the field. (Pillai and Bhat, 1987).

A method for rearing the carabid predator, *Calleida splendidula* (F) in the laboratory is described. Short descriptions of the adult and immature stagesof the beetle and feeding potential of the predator on *Opisina arenosella* wlk., caterpillars are also furnished. The adult beetle, on an average, consumed on fully grown caterpillar of *O. arenosella* in every three days. Eleven to thirteen second instar caterpillars of *O. arenosella* were consumed by each grub during the entire grub phase. The adult beetle lived for 6 to 14 months in laboratory cages. A comparison between *C. splendidula* and *Parena nigrolineata* (choudoir) is also is furnished. (Pillai and Nair, 1990).

Biology, habits, field activity and abundance of the predator *parena laticincta* Bates were discussed. (Rao *et.al.*, 1978).

Larvae of *Anklopteryx octopunctata candida* were observed to prey on eggs and early instar larvae of the Xyloryctid coconut pest, *Opisina arenosella*, in Kerala, in 1982. The adults of *A. octopunctata* are described. They are nocturnal and during day they rest on leaflets of the coconut palm. Feeding trials revealed that each first instar larva consumed 2 - 3 eggs of *O. arenosella*/ day. Females of *A. octopunctata* laid pedunculate eggs in batches on the infested coconut leaflets, which hatched in about two days. The larval period ranged from 9 - 11 days and there were three larval instars. Pupation occurred in or near the leaf axils, and the pupal period ranged from 10 - 12 days. The average life span of *A. octopunctata* reared on honey in the laboratory was 50 to 72 days for the female and 17 to 35 days for the male. The predator appeared in the field from April to October, and the population peaked in June and July. (Sathiamma *et.al.*, 1985).

Larval Instars : Detailed laboratory studies at ambient conditions of 23° to 32° C and 75 to 90% RH at 12 : 12 photoperid show that the insect *Opisina arenosella* has eight larval instars; the ratio of head capsule width of successive instars is more or less constant and ranging between 1.2 and 1.7. The graph showing log head capsule width and instar is a straight line. (Santhosh Babu and Prabhu, 1987).

Parasites : Trichospilus pupivora Ferr. is one of the important parasites of *Nephantis serinopa* Meyr., a serious caterpillar pest of coconut palms. Behavioural aspects of coartship, mating and oviposition of *T. pupivora* are described. The coartship activities involved the characteristics movements of antennae, head and abdomen. Usually, mating occurs while the insect are within the host pupae. Oviposition take place within two to four hour after emergence. The females lay eggs inside the host pupae by piercing the pupal skin with their ovipositor. More than one females are found to oviposit at a time on the same pupa. (Beena, 1980).

Meterridea hutsoni (Nixon), which is recently recorded larval-pupal braconid parasite of *Opisina arenosella* walk., has a distriution limited to certain areas of Malabar in Kerala. Observations on certain aspects of the biology and behaviour of this parasite are given. Females live longer than males with 50% honey, females lived for 12.4 days whereas males live for 9.3 days. In the field, males out number females, their ratio being 4:3. Total duration of life stages from egg ot adult varies from 17 to 23 days (Ghose, and Abdurahinman, 1984).

Investigations carried out in Thikkodi (Kerala) have shown that *Apanteles taragamae* is an efficient biocontrol agent of *Opisina arenosella*. The mean percentage of its parasitism on *O. arenosella* is eight in the study area for 1983-84. Field data on the rate of parasitism idicates its promising potentiality as a biocontrol agent of the pest. The efficiency of this parasitoid is assessed in comparison to the other larval parasitoides of *O. arenosella*. (Ghose and Abdurahiman, 1985(a)).

Studies on the intensity of successful parasitism and development of progeny of *parasierola nephantidis* reared on different instars of *O. arenosella* caterpillars revealed that the stage of the hot larvae used was critical in determining the acceptance by the parasitoid, number of eggs laid, number of off spring developed and the proportion of males in the brood. Fifth and early sixth instar caterpillars of *O. arenosella* were ideal stage of the host for large scale multiplication of *P. nephantidis* in the laboratory. The parasitoid was found to be capable of suppressing the early instar caterpillar of the pest by stinging, even though they were not ideal stages for the development of a good number of parasitoides. (Pillai and Bhat, 1986).

Attempts at laboratory mass-rearing of the parasite *Parasierola nephantidis* mues resulted in the production of females only. An investigation was therefore made to see whether unmated females *Parasierola nephantidis* could parasitise *N. serinopa* to a useful degree. Females reared in the laboratory were confined with host larvae either before or after being allowed to mate. Unmated females laid eggs on the host and gave rise to male only, whereas mated females produced progeny of both sexes; both mated and unmated females had 3 reproductive cycles lasting 12-13 days each, with intervals of 4-5 days between. The successful parasitisation of hosts with male eggs by unmated females of *P. nephantidis* suggests that the latter can be used for biological control of *N. serinopa*. (Murthy, and Santhanakrishnan, 1978).

Evaluation of the intensity of natural parasitisation of *N. serinopa* pupa in Quilon and Allepey districts of Kerala during 1975 and 1980 revealed that more than 50% of natural pest suppression was due to the pupal parasitoids of which the chalcidids *Brachymeria* spp. were the major ones, and among them, *B. nosatoi* was the most dominant species followed by *B. nephatidis*. Activity of *B. nosatoi* was more during summer season, which is also the peak period of the the pest. The intensity of natural parasitisation by *Trichospilus pupivora* was meagre. (Pillai and Ramachandran Nair, 1981).

Laboratory and field observations carried out in India showed that part of the population of *Trichosilus pupivora* Ferr. a gregarious pupal parasite of the coconut pest *Opisina arenosella* Wlk. (*Nephantis serinopa* Meyr.) aestivated in pupal stage between March and May. The poor searching and dispersal ability of the parasit, its low tolerance to non-optional temperature, inability to discriminate between parasitised and unparasitised hosts leading to superparasitism and inability to compete with other pupal parasites were the major factors responsible for its inefficiency and low intensity of parasitism in the field. (Pillai and Ramachandran Nair, 1982 (a)).

A Technique to mass rear the chalcidid *Brachymeria nosatoi* Habu, a solitary pupal parasite of the coconut pest *Opisina arnosella* wlk (*Nephatis serinopa* Meyr), by simulating field conditions is described. The chalcid oviposited in pupal cocoons that had been kept intact inside silken galleries. Other species of brachymeria, such as *B. nephatidis, B. himeatteave* and *B. lausus* (wlk) could also be reared in the laboratory, using this technique. Newly extdysed pupae were accepted for parasitism as well as those at an advanced stage of developement. (Pillai and Ramachandran Nair, 1982(b)).

The biology, parasitic behaviour, and intensity of natural parasitism of *Bracon hebetor* (*Microbracon hebetor/Habrobracon juglandis*) were studied. The occurrence of *B. hebetor* as a natural parasitoid of *O. arenosella* from different tracts indicates its wide distribution and its potential in the natural supression of the pest. The long life span, long ovipostion period, high fecundity and host finding are the major attributes which make *B. hebetor* an important bilogical control agent of *O. arenosella* on coconut (Sathiamma, *et.al.*, 1986).

Bracon brevicornis wesmael (Hymenoptera : Braconidae) is another important indigenous larval parasitoid of *O. arenosella*. An account of the bilogy and morphology of the immature stages of this parasitic wasp is given. There are five larval instars and the duration of developement from egg to adult takes about 7 to 8.5 days. (Sudheendra Kumar *et.al.*, 1982).

Laboratory studies on the control of *Opisina arenosella* by micro-organisms were conducted. Of the three micro-organisms tested, *Aspergillus flavus* L. and *Serratia marcescens* Bizio caused the greatest mortality while *Bacillus thuringiensis* Beriliner was not effective in causing caterpillar mortality. Pathological observation are reported. Re-isolations of the respective pathogens were successfully made for the dead caterpillar and also from the incomplete pupae. (Muthukrishnan and Rangarajan, 1976).

Hyperparasites : Some parasites which attack the serious pest *Opisina arenosella* (*nephantis serinopa*) in turn serve as host to some hyper-parasites reducing the efficiency of biological control.

Over half the cocoons of the parasitoid *Apanteles taragamae* were parasitised by the hyperparasites, with *Aphanogmus manilae* accounting for 27.1% followed by *Meteoridea hutsoni* (17.4%). *Pediobius imbreus* also caused 29.8% parasitism on *Bracon breviconrnis*. Brief description of the biology of some of the hyperparasites also given. (Ghosh and Abdurahiman, 1985(b)).

Opisina arenosella is attacked by more than 29 parasites and 10 predators. Among the larval parasites, *Apanteles taragamae* outstands in field population. However, its efficacy is seriously imparied by the activity of several hyperparasites. *Goniozus nephantidis* and *Elasmus nephantidis* occur only in very low proportions in Malabar. *Meteoridea hutsoni* is restricted to certain regions of Malabar, where its percentage is surprisingly high. Among the pupal parasites, *Trichospilus pupivora, Brachymeria nephantidis* and *B. nosatoi* show a fluctuating rate, while all other chalcid parasites exhibited a very low rate (Ghosh and Abdurahiman, 1987).

The parasitoid complex of *Opisina arenosella* showed the occurrence of many potential hyperparasitoides, attacking the different stages of the primary parasitoid. Among them are, *Pediobious imbreus, Aphanogmus manilae, Erytoma braconidis, E. albotibialis, Brachymeria nephantidis, Meteoridea hutsoni* and *Elasmus nephantidis*, which have been found to parasitise the primay Braconid parasitoids. Brief account of the biology of these hyperparasitoids and their role in reducing the population of the primary parasitoids are given (Ghosh and Abdurahiman, 1988).

Perisierola nephatidis M. and *Bracon brevicornis* W. are the effective larval parasites of coconut leaf eating caterpilar, *Nephantis serinopa* Meyr. in Mahuva area of Gujarat State. The cocoons of both larval parasites collected from the coconut field are found to be parasitized by two species of *Pediobius*, identified as *bruchicida* (Ronadami) and *imbreus* (Walker). (Kapadia, 1983).

Chemical insecitcides : Since DDT is harmful to parasites and predators of *Nephantis serinopa* Meyr. trials were carried out to select a suitable insecticide, DDT 0.2% Carbaryl 0.2% BHC 0.2%, Quinalphos 0.05%, methyl parathion 0.05%, endosulphan 0.07% and monocrotophos 0.05% were sprayed at the rate of 6 1/10 yar old palm and the result were recorded in each case. All these insecticides proved to be effective but ony BHC, mehtyl parathion and manocrotophos were recommended due to their reduced toxicity to useful insects (Kapadia, 1982).

Phosalone was the most (highly) toxic insectide against *N. serinopa* and ***B. breicornis***, with LD50 value of 0.024, followed by endosulfan, malathion and quinalphos, with LD50 values of 0.024, 0.052 and 0.057 respectively. Fenthion, phosphamidon, monocrotophos, fenitrothion DDVP, BHC and carbaryl were much less toxic and were not useful for controlling the caterpillar pest. All the insecitcides except phosalone, quinalphos, manocrotophoss, carbaryl and BHC showed negligible residual toxicity by the eighth day of insecicide application, on the 16th day none of te insecitcides except carbaryl and monocrotophos had any residual toxicity to the larvae and the residues of monocrotophos and carbaryl still had substantial toxicity (Mathew *et.al.*, 1979).

On the seventh day after application, dichlorvos (0.02%), BHC (HCH) (0.2%) or trichlorphon 10.06%) reduced *O. arenosella* infestation on coconut palms. (Ponnamma, 1984).

Chemical control of *Nephantis serinopa* poses serious limitation and spray recommendations given so far are somewhat difficult with tall coconut palms. A new method of administering of systemic insecticides through root adopted at the parasite Breeding Station of Ambajipet and was found effective against *N. serinopa.* This method obviates harmful effects against beneficial parasites and predators and avoids hazards to the operator. (Rao *et.al.,* 1980).

Control Measures : Coconut trees of all ages are attacked. Biological control relying a larval parasitoid, the bethylid, *Parasierola nephantidis* is the most effective in controlling the pest if used during the initial period of infestation (less than two larvae per leaf). The optimum level of release is 1 : 8 of host-parasite ratio. The parasite should be released under the coconut trees when the pest is in the second or third instar larval stage. Parasite release trap may be used to release the parasite at the site of feeding. Parasites should not be released in the crown region since they will be killed by predators like spiders and reduvids. Second release can be made three months after the first release.

Release the larval (Bethylid, Braconid an Iohneumomid) and Pupal (Eulophid) parsites and predators periodically after January to check the build up of the pest during summer.

Remove and burn all affected leaves and leaflets.

Under high pest pressure (more than 2 larvae/leaf) chemical control become inevitable. Spray Malthion 0.05% to cover the undersurface of the leaves thoroughly in case of severe epidemic outbreak of the pest in young palms. A better management was developed keeping the safety of the natural enemies in mind.

Root Feeding with Systemic Insecticide : Select a fresh and live root, cut sharply at an angle and insert the root in the insecticidal solution, cotaining monocrotophos 36 WSC 10 ml + water 10 ml in a 7.5 x 10 cm polythene bag. Secure the bag tightly to the root with a cotton thread. Twenty four hours later, check whether there is absorption. If there is no absorption select another root. These methods should not be rsorted to as a routine practice and it is suggested only for cases of severe epidemic outbreak of the pest and when the survival of the tree is threatened.

4. Coconut Scale

The biology of coconut scale (*Aspidiotus destructor* sign) was studied in pot culture on tall and dwarf coconut types. For both sexes the incubation period of the egg was 8 to 9 days. The females had two moulting stages while the males usually had three with a total life cycle of 34 days for the females and 29 days for males. On the dwarf coconut the insects were smaller, which had longer life cycles and higher mortality (39.1%) than on the tall coconut types (28.4%). (Jalaluddin and Mohanasundrama, 1989(a)).

The scale insect population has a significant correlation with maximum temperature, sunshine hours and negatively correlated with morning relative humdity. Two aphelinid parasites, viz., *Aphytis chrysomaphali* Marcat and *Cauca parvipennis* Gahn., were collected from field collections and they were active mostly in the months of May and July. Three predators are also reported to be active. (Jalaluddin and Mohanasundaram, 1989(b)).

A new species of *Pseudoscymnus* that was discovered preying on *Aspidiotus destructor* Sign., infesting the lower surface of the leaves of coconuts palm in several locatlities of Karnataka and Andhra Pradesh in 1968-72, is described from the adults as *Pseudoscymnus dwipakalpa*, sp, n. This is the first record of the genus from India, a key is given to these species (Ghorpade, 1977).

Eleven species of coccinellids, all of which prey on *Aspidiotus destructor* sign. are reportd from the silken nests of spiders on coconut trees around Bangalore (Ghorpade, 1979).

Occurrence of five new species of scale insects in Gujarat is reported. Characters of scale, intensity of damage, variety and part of palm affected and the parasites/predators recommended for the biological control with respect to each of these species are discussed (Kappadia, 1984).

5. White Grub

Leucopholis coneophora Burn. (Melolonthinae : Scarabaeidae : Coleoptera). Cockchafer, white grub *Leucopholis coneophora* Burm, feed on the roots of coconut palms and is a severe pest in Kerala and Karnataka States. The incubation, larval and pupal period averaged 23,270, and 26 days respectively. Adults emerged from the soil to mate at dusk and returned to soil within 25 - 30 minutes. (Abraham and Mohandas, 1988(a)).

Chemical Control : Four chlorinated hydrocarbon insecticides, HCH, heptachlor, aldrin and chlordane were evaluated for control of the immature stages of the white grub using bio-assay techniqes. Aldrin and heptachlor were the two most active compounds, weight for weight, against second and third instar grubs, but a projected cost performance basis HCH and heptachlor emerged as the most promising candidates. Field evaluation of these two insecticides under controlled conditions proved that *L. coneophora* could best be controlled with hepotachlor appled at 1.4 Kg a.i./ha in the month of June or with HCH applied twice at 5 Kg a.i./ha in June and September. (Abraham and Mohandas, 1988(b)).

Experiments conducted at Kayankulam (Kerala) have shown that cassava and coconut were favourabe hosts of *L. coneophora.* A higher build up of pest population is, hence, likely in coconut gardens in which cassava is grown as an intercrop.

Influence of weather on the larval populations of cockchafer beetles in coconut was studies at Regional Agricultural Research Station, Pilicode (KAU). Results indicated that the abundance of white grub in the root environment of coconut palms was under the influence of mean monthly rainfal and was negatively correlated with mean maximum temperature. The populations were higher during June and July than during February to May. The top 80 cm of the soil harboured the maximum number of gurbs. It is advocated that prophylactic insecticidal treatment of the top layer of soil during the period of heavy rainfall will aid reducing the pest build up in endemic areas. (Ranjith *et.al.*, 1988).

In Karnataka, three species of white grubs damage coconut. While *Leucopholis coneopora* Burm. and *L. lepidophora blanchard* confine themselves to heavy rainfall areas of Coastal Karnataka and Malnad. *Holotrichia serrata* Fabricus is endemic in the plains. Ecology and chemical control of *H. serrata* were studied at UAS, Bangalore. Five soil insecticide including chlordane 20EC, dursban 20.0 EC, phorate 10G, counter 5G and temik 10G, were tested under irrigated and unirrigated conditions at two levels each. Dursban 20.0 EC at 10 ml, and counter 5G and phorate 10G, at 20g/palm with irrigation after insecticide treatment gave good control of grubs. (Veeresh, 1983).

6. Bagworm

Occasional sudden outbreak of bagworm (*Monatha albipes* M. Psychidae : Lepidoptera) occurs in coconut palm. The bagworm caterpillar is dirty brown with dull white and brown mottlings on its head and thorax regions. The caterpillar feeds on the green portion of leaf frond resulting in numerous holes on the coconut leaves. This reduces the photosynthetic activity and consequently the general vigour of the palm. Full grown catepillars measure 1.5 to 2.0 cm in length. The worms concentrate in the fronds of bottom whorl leaves in the palm crown and cause severe damage.

Management practicces of the pest include cutting and burning the leaves. Spraying the leaves with dichlorvos or DDVP at 2.0 ml in one litre of water with 0.5 ml liquid soap with the help of a high volume sprayer to the point of runoff is also effective (Sadakatulla *et.al.*, 1997).

7. Termites

Termite (*Odontotermes obesus* Ramb) infestation on coconut is more common in nurseries. Germinating nuts and transplanted seedlings are frequently attacked. Various termite species attacking coconut palm have been tabulated and their harmful effect on yield and control measures were reviewed. (Kashyap *et.al.*, 1984).

The *in vivo* experiments showed the significant superiority of spray with aldrin 0.15% of HCH 0.25% or chlordane 0.1% on the base and upto 2 metres height of the trunk of coconut palm in control of *Odontotermes obesus* and aldrin 0.15% or chlordane 0.1% on the plaited coconut leaves even when the plaited leaves were kept on the bare ground having continuous occurrence of this subterranean pest, (Sadakatulla and Ramachandran, 1990(a)).

8. Mealy Bug

Mealy bug is one of the insect pests responsible for nut fall in coconut, right from button stage upto mature nuts. The need for a detailed study on the infestation of the pest with special emphasis in the extent of the damage on unopened spathe and loss of yield is highlighted. The reason for the co-existence of the coreid and mealy bug are to be studied. It is quite common that healthy trees with severe infestation of mealy bugs have even five barren bunches on their crown. Infestation on spathe affects the vigour of the inflorescene. This pest could be controlled for a short period by spraying with dimethoate. (Radhakrishnan, 1987).

The nature and extent of damage caused by mealy bugs on coconut is reported. A control measure with 0.2% carbaryl spray is recommended at 2 - 3 weeks interval. (John, 1988).

9. Coreid Bug

Paradasynus rostratus Dist., a coried bug was recently recorded as a new pest of coconut in India. The nymphs and adult feed on the tender nuts causing immature nutfall and malformation. There are five nymphal instars and the total life period is 82-97 days. The description of various stages, life history, nature and extent of damage and the distribution of the pest in space and time are discussed (Kurian *et.al.*, 1979).

Three insecticides viz., Carbaryl, HCH (both at 0.1 and 0.2%) and endosulfan (at 0.05 and 0.1%) were tested for the control of the coreid (*Paradasynus rostratus*) on coconuts in Kerala in 1982-83. Effective control was achieved with 0.1% carbaryl, 0.2% HCH or 0.1% endosulfan (Visalakshi *et.al.*, 1987).

10. Scolytid Bark Borer Beetles

Damage caused to coconut palms by *Xyleborus crassiusculus* (Matsch) (semiopacus Eich) in Karnataka is described. Bore holes were found in the trunk at heights upto 140 cm. The trunks of infested palms shrunk and infested palms had an average of 9.5

fronds, as compared with 19 on uninfested palms. Considerable number of the nuts on infested palms fell prematurely. when infested trunks were opened, severe tunnelling was found to a depth of 18 - 22 cm and all stages in the life-history were present in the tunnels simultaneously. (Deshpande *et.al.*, 1982). This can be controlled by (i) stem injection through a stove wick soacked in 0.2% Fenthion or 0.2% Dichlorvos and plugging the hole and repeating the treatment using the same wick and hole a month after and (ii) swab the palm surface with BHC 0.1%.

11. Weevil Diocalandra

Diocalandra stigmaticollis has been fond as pest in the coconut gardens of Kanyakumari, Tirunelveli, Thanjavur and Coimbatore districts in Tamil Nadu affecting the bark of the young and grown up palms. The adult weevil has a long, pointed snout with a gentle inward curve. It is about 6 mm in length, and 1.5 mm in width. The larvae is dull white, about 8 mm in length, having a wrinkled and curved body without legs. Posterior half of the abdomen is well developed. The larvae burrow in to the soft tissues of the bark and never found deep in the fibrous portion of the stem. The cylindrical cavities of the adult weevils are filled with resinous secretions, similar to thick molasses. The symptom of the attack is very often mistaken for early stages of stem bleeding. In the affected palm, the production of leaf and spathe is delayed. The buttons shed considerably and setting affected upto 75%. the yield of the palm is affected completely and the palm succumbs to the attack in two or three years (Louis and Chandrasekharan, 1980).

B. Mites

The coconut eriophyid mite *Eriophyes guerreronis* was first identified in Mexico during 1965. The mite is minute (200-250 μ in size) with two pairs of legs. It is pale and worm like. The life cycle of this Aceria group consists of egg, two larval instars and adult stage. The total life cycle lasts for 10-15 days.

They are seen in colonies in the perianth of developing nuts and live in white tender portion covered by the inner bracts of perianth and suck the sap from the tender meristematic mesocarp tissues. Initially they inhabit the floral bract and the tender region just below the perianth. Early stage of attack of this mite resutls in dropping of buttons. Continuously sucking of the sap ultimately results in formation of brown patches. As the nut grow they wilt and longitudinal fissures develop on the outer surface of the nuts. Severe infestation leads to poor nut growth, reduced copra content and cracking and hardening of husk. Maturing nuts upto nine months of age harbour the mites.

Control Measures : As these mites multiply in large numbers in short span of time prompt quarantine measures are suggested to prevent the transport of infested nuts. Monocrotophos at 0.04 percent or dimethoate 0.03 percent with one ml of liquid soap per litre of spray solution may be sprayed on spathes, buttons and tender nut twice at one month interval using a high volume sprayer to the point of run-off. Adoption of root feeding of monocrotophos 10 ml + water 10 ml per palm twice at 4 days interval following due precautions may effectively control the mite. The solution must be put in to 100 ml capacity polythene bag and the selcted cut end of the root must be inserted in to the bottom of the polythene bag and the mouth of the bag has to be tied. Mature nuts must be harvested before treatment and the nuts should be harvested and consumed only 45 days after treatment sicne the chemical residual effect persists. (Sadakatulla *et.al.,* 1999).

C. Nematodes

Burrowing nematode : Radopholus similis infestation on coconut, arecanut, banana and pepper was reported from various districts of Kerala, Karnataka and Tamil Nadu. *R. similis* occurred maximum in sandy-loam soil. Banana as an intercrop in coconut or arecanut was found to favour multiplication of this pest. The semihard orange coloured portion of the roots of coconut harbours the highest number of nematodes. Peeling and slcing of roots during processing resulted in maximum nematode recovery. Optimum temperature range for extraction of *R. similis* from coconut roots is 4.0 to 14.0° C. *R. similis* produced small elongated reddish brown cortical lessions on roots of coconut and arecanut which later coalesce and cause extensive root rotting. Population from coconut, arecanut and banana did not differ in their pathogenicity. The coconut cultivars, West Coast Tall, Dward Orange, Dwarf

Green, Gangabondam, Tall x Dwarf, Dwarf x Tall, and Tall x Gangabondam were susceptible to the burrowing nematode. (Koshy *et.al.*, 1975).

Java Tall (from Indonesia), Klapawangi (from Malaysia) and Andaman Giant and Kenthali (from India) were considered fairly tolerant to *R. similis.* Java Tall alone or in hybrid combinations with Malaysian Dwarf Yellow were recommended for cultivation in *R. similis* infested areas. (Sosamma *et.al.*, 1988).

Systemic nematicides, aldicarb, carbofuran, phorate and phenamiphos (fenamiphos) at 10, 15 and 25 kg a. i/ha applied 3 times at the same rate for one year were evaluated for the control of *R. similies* in coconut nurseries. All treated plants showed an increase in growth with aldicarb giving the best plant growth. Except 10 kg a.i./ha of aldicarb and 10 and 25 kg a.i/ha of carbofuran, all dosages and nematicides reduced nematode populations. Complete control of *R. similis* was obtained with phenamiphos/phorate at 25 kg a.i./ha. (Koshy *et.al.*, 1985).

D. Rodent Pest

Rat : Rat is serious pest of coconut in Lakshadweep, Andhra Pradesh, Kerala and Tamil Nadu. The effectiveness of two traps, bambo snap back-cum-nose and plank trap death fall trap, was tested in coconut plantation (23.5 ha area) for 24 months and 9 months respectively. The former gave 15.5% catch of *Bandicota bengalensis* and latter gave 24.6 catch of *B. indica, Rattus rattus* and *Tatera indica.* (Kurian *et.al.*, 1977).

Wooden traps and sherman metal traps were found effective in capturing adult and sub-sdult rats respectively, while wire cages were comparatively less successful. With the increse in palm density/ha, a rise in rodent infestation levels and relative damages were observed. Ten times more rodents were trapped on the crowns of coconuts than at ground level, suggesting to undertake poison and trapping operations on the crowns of coconut palms rather than at ground level. (Advani, Ranjan, 1984).

The bait preferences of *Rattus rattus* were studies using commonly available baits with additives. It was observed that the split forms of raw rice mixed with 2% sugar/ jaggery was the best bait to mix poison for the control of this pest. (Bhat and Sujatha, 1989(a)).

One time and two times baiting with multiple dose anticoagulants have reduced the population of black rats by 50% and 70% respectively. Continuous baiting for 3 times has reduced the population by 94% and damage by cent percent. The cost involved for 3 times baiting was not much when compared to the returns from the operation. (Bhat and Sujatha, 1988).

Warfarin and fumarin are the two anticoagulant rodenticides available in Inida for rodent control operations. The toxicity and relative acceptability of these poisions of the black rat, *Rattus rattus* wroughtoni Hinton,and important rodent pest of coconut and coccoa in South India were evaluated in the laboratory. There was no significant difference in the toxicity between these poisons (LFP 50 around 4 days for both the posions) to *Rattus rattus,* but there was much difference in their acceptance. After consuming poison baits, the animal started dying from the third day onwards. The peak moratality was on the fifth day for warfarin and seventh day for fumarin. Cent percent mortality was achieved within 12 to 14 days feeding. However, the baits mixed with fumarin were preferred significantly to those mixed with warfarin and hence, recommended to use such poison bait continuously for two weeks in the field. (Bhat and Sujatha, 1989(b)).

Investigations on crown baiting of coconut palms using warfarin and bromadiolone baits embedded in paraffin have revealed that both are potent anticoagulants bringing a minimum mortality rate of 84% rodent population. (Rao *et.al.,* 1983).

Damage to tender coconuts by black rat (*Rattus rattus*) in India is reported to range form 9 to 15% in Andhra Pradesh to 35 and 50% in Andaman and Lakshdweep islands, respectivey. The species nests in the palms and rarely comes down to the ground for foraging. Damage other than to tender nuts is also described, and control measures are outline. Baiting twice at 12 days intervals with 10g bromadiolone wax cakes placed in the crown, on every fifth palm gave good control of the pest (Bhat and Sujatha, 1989 (c)).

The LD 50 valur wth 95% confidence limits of bromadiolone to *Rattus rattus* was found to be 0.51 mg/kg body weight. One and two days exposure of rats to this poison resulted in 93% and 100% mortality respectively. In coconut fields, application of 10g bromadiolone baits two times, at an interval of 12 days on the crown of every fifth palm completely reduced the rat damage. (Bhat and Sujatha, 1991).

Bandicota bengalensis damages coconut seedligns by eating the roots and growing into the husk and shells, while *Rattus rattus* damages the nuts on the palms. Observations revealed that trunk banding alone does not control rat damage; on the other hand, it may even confine rats to particular palms. Such rats cause more damage than the rats on unbanded palms. Isolated rats could be successfully controlled by zinc phosphide treatment. Trunk banding followed by poison baiting effectively controlled *R. rattus*. Control of *B. bengalensis* in nurseries with Zinc phosphide baiting followed by fumigation with aluminium phosphide proved efficient and profitable (Guruprasad and Srihari, 1983).

II. Diseases

The coconut palm is as much prone to diseases as any other crop. Fatal disease like Thanjavur wilt, and debilitating diseases like root (wilt), stem bleeding, bud rot (*Phytophthora palmivora*), leaf blight (*Pestalotia palmarum*), button shedding, and nutritional disorders including tapering wilt occur in various parts of the country.

1. Thanjavur wilt

Among the several diseases affecting coconut, Thanjavur wilt is the most serious and devastating dissease. The symptoms of the disease consists of exudation reddish brown viscous fluid from the basal portions of the stem accompanied by flaccidity of leaves. Th bleeding patches begin from the base and extend upwards as the disease progress. Some palms showed only wilting unaccompanied by bleeding. The leave become brown, droop down and fall starting from the outer whorl, leaving the spindle alone or in extreme cases even the spindle is blown off leaving the decapitated stumps. Button and premature nut fall is common and the yield reduction will be substantial after increase in yield in the initial stage of the disease. (Ramanathan, 1987). *Ganoderma lucidum* (Leys) korst is found associated with the disease in most of the cases and is considered as the main causal agent of the disease.

Among the various plant extracts tested, banana rhizome extract and 'kolingi' root extract were highly inhibitory to the growth of *Ganoderma lucidum* giving 1.2 cm and 4.0 cm radial growth respectively as against 8.6 cm growth in potato dextrose agar medium without plant extracts. (Bhaskaran *et.al.*, 1987).

Trichoderma harzianum and two isolates of *T. viride* were found to be antagonistic to *Ganoderma lucidum* (Leys) Karst., the fungus associated with Thanjavur wilt of coconut. *T. harzianum* lysed the hyphae of *G. lucidum* while the two isolates of *T. viride,* besides lysis also punctured the host hyphae, thus, causing physical parasitism also. Neem cake extract completely inhibited the growth of *G. lucidum* and the extract from banana rhizome and 'Kolingi' roots also inhibited the growth significantly. Neem cake encouraged the population of saprophytic soil micro flora especially *Trichoderma* and was effective in the control of Thanjavur wilt disease. (Bhaskaran *et.al.*, 1988).

An attempt was made to study the internal pattern of tissue damage, through destructive sampling, in Thanjavur wilt affected palms, the etiology of which is uncertain. Results have shown that the degree of infestation in the trunk region was more and a decreasing trend was noticed towards the crown region. The vascular region was completely disintegrated in an adult palm while in the young palm, necrotic areas were observed upto two-thirds of the total height. Isolation from infected roots and stem tissues yield the fungi, *Botryodiploidia theobromae* pat., *Ganoderma lucidum* (Leys) karst., *Fusarium solani* and some sterile mycelia. The carbohydrate content was found to be lower in the diseased roorts comapred to healthy. The lower content of carbohydrates in diseased roots suggest that the infection may primarily be incited by a root invading pathogen (Ramadoss *et.al.*, 1986(b)).

Ganoderma lucidum was isolated from the roots of diseased palms, which after artificial inocultion caused root rot in healthy coconuts. An increase in soil temperature increasd the severity of the disease and an increase in soil moisture decreased the severity.

Incorporation of organic manure in the soil in coconut basins (especially neem cake) and irrigation during summer months reducedthe severity of the disease. Soil drenching with 1% Bordeaux mixute at 40 litres/palm and root feeding with 2g of Aureofungin-soil with 1 g of copper sulphate in 10 ml of water, both done 3 times at quarterly intervals preferably during July, October and January were very effective in co ntroling the disease (Bhaskaran *et.al.*, 1990).

Application of carbendazim or carboxin through roots at the rate fo 2 g in 100 ml of water thrice a year was found to be equally effective as that of soil drenching with 1% Boredeaux mixuture and stem injection of Aureofungin-so1 2g plus copper sulphate 1g in containing the disease. (Ramadoss *et.al.*, 1986(a)).

Other disease management practices include, intercropping of banana in coconut, soil application of 500g in each of magensium sulphate and copper sulphate and treatment with 500g inoculum of *Trichoderma harzianum* multiplied in 50 kg cattle manure.

However, these control measures will be effective only if the palms are treated during the early stages of infection. By the time clear visible symptoms beign to appear on the palms much damage has already been done.

Hence studies were conducted at the Coconut Research Station, Veppankulam (Tamil Nadu) to detect the disease before the expression of external symptoms. Laboratory studies followed by field testing of chemo-diagnostic method were carried out. The findings revealed that EDTA and TTC tests are useful for early diagnosis of the disease. Root sampling procedure for the diagnostic tests has also been standardised for reliable results.

In EDTA test, one gram of chopped root tissue is immersed in 10ml of 0.3 methylene diamine tetra acetic acid (EDTA) (disodium salt). The optical density values are then recorded after one hour of incubation.

In TTC test, one gram of chopped root tissue is imersed in 10 ml of 0.1% 2,3,5 triphenyl tetrazolium chloride (TTC) solution for 24 hours. After incubation, the optical density values were measured.

Experimental results indicated that root sampling from any direction in the basin area of the palm at 15 to 30 cm depth is optimum for EDTA and TTC tests. Root samples can be stored at room temperature for 6 hours and in a freezer at 4 degrees for 18 hours for more reliable results. Periodical testing of the palms revealed the critical OD values as 0.25 to 0.62 in EDTA test and 0.14 to 0.42 in TTC test. This forecast becomes available 4 to 14 months before the expression of visual symptoms of the wilt disease in the field. This will be useful to detect the disease in advance and to take up appropriate control measures (Karthikeyan, 1998).

2. Root (Wilt) Disease

Root (Wilt) is a debilitating diseases of coconut palms in Kottayam, Allepy, and Quilon districts of Kerala. It has been noticed in isolated areas in Tirunelveli and Kanyakumari districts of Tamil Nadu.

Flaccidity of leaflets is the earliest foliar symptom of the disease. Paling, yellowing, ribbing and bending of the leaflets are the other typical foliar symptoms. Diseased palms also exhibit extensive root damage and rotting preceeeds from the tip downwards. Progress of the disease results in stunting of the crown due to a reduction in the number of leaves. Successive leaves become shorter. Young palms if infected before flowering, may not flower at all or will have delayed flowering. In adult trees, nut yield may be reduced progressively year after year.

The predisposing factors are soil nutrient imbalance with mineral deficiencies, low pH and reduced soil microbial activity. The disease is pronounced in ill drained and poorly aerated soils. The soils in endemic areas have recorded low available potassium, exchangeable calcium and magnesium ions.

Studies conducted over years have ruled out the involvement of fungi, bactria, nematodes or viruses as causative agents though they are frequently associated with diseased palms. However, constant association of micoplasma like organisms (MLOs) in the tissue of young leaves has been found in diseased trees. The lace wig bug *Stephanitis typica* (Distant) acts as a vector transmitting pathogen to othe trees (Sankaralingam and Usman, 1996).

The presence of Mycoplasma like organisms was identified in seive tubes of roots, tender stem, petiole, and leaf bases of all the disesased palms examined as against their total absence in healthy palms. Tranmission experiment in insect proof cages using healthy coconut seedlings showed that lace wing bug (*Styphantitis typica*) is the vector. MLOs were observed in brain and salivary glands of lace wing bug given a A+1 period from 18 to 23 days. MLOs were also observed with salivary galnds of plant hopper (*Proutista moesta*) with a A+1 of 40 days. MLOs were transmitted from diseased coconut palms to periwinkle through dodder. Remission of symptoms of the disease also could be obtained by Oxytetra cycline (OTC) treatment of diseased palms. (Solomon and Govindan Kutty, 1991).

An attempt has been made to check the foliar yellowing by applying different nutrients and to study its effects on the growth and yield of palms. Foliar sprays with 2% Mg SO_4 and 1% Fe SO_4 and basal application of 500g MgO reduced foliar yellowing by 26% in root (wilt) affected palms and 87% in healthy palms. In diseased palms the applications increased yield by 106%, while in healthy palms, the increase was 95%. (Varkey *et.al.*, 1979).

As infected trees serve as a source of contamination, such trees have to be cut and burnt. Waterlogging has to be prevented and to facilitate proper aeration to roots, the interspace in groves has to be ploughted twice a year. Regular irrigation has to be given during drought period. To control vector (lace wing bug), malathion @ 0.1% has to be sprayed. As occurrence of leaf-rot disease is common in root (wilt) affected palms, regular and sequntial spraying with bordeaux mixture (1.0%), Dithane M. 45 (0.3%) and fytolan (0.25%) has to be followed to reduce leaf-rot incidence.

Results of investigations have shown that improvements of soil physical conditons, and adoption of inter and mixed cropping system, use of better planting materials and adoption of package of practices gave higher yield inspite of the diseased condition.

3. Stem Bleeding

Two distinct types of symptoms were observed in coconut palms affected by the stem bleeding disease. In the first, bleeding was sparce and watery; crown leaves become desiccated and nuts were shed while the centre of the trunk rotted, resulting in tree death in about a year. In the second type, bleeding was red and turbid and the ooze usually hardened with no trunk rotting; trees recovered in six to eight months. Certain fungi were isolated consistently from rotted tissue beneath bleeding patches but not in the absence of rotting (Sulladmath, *et.al.*, 1980).

Experimental evidences have shown that the causal organism of this disease is *Thielaviopsis paradoxa.* Sample of diseased tissues were collected from coconut palms affected by stem bleeding disease, growing in a number of locations in Karnataka, Kerala and Goa. The fungus *Thielaviopsis paradoxa* was consistently isolated from all locations. Characteristic symptoms of stem bleeding, in the form of rusty brown discolouration of bark, were reproduced in all the eleven inoculated palms located in Kasaragod, Apangala and Vittal. Inculum comprised of small pieces of coconut rachilla, used as a substrate, which were completely covered by an actively sporulating mycelium of *T. paradoxa.* These bits were introduced in bore-holes made on palm trunks. Symptoms appeared at inoculated sites within two wks (at Appangala) to eight wks (Kasaragod and Vittal). The organism was reisolated from the tissue gathered from the inoculated palms. (Nambiar *et.al.*, 1986).

In vitro interaction studies revealed that *Trichoderma Viride* exerted maximum inhibition of *T. paradoxa* (90%) followed by *T. harzianum* (86.6%) and *Aspergillus niger* (50%) (Gowda, 1987).

A method for indexing the disease severity in stem bleeding affected coconut palms, based on lesion size and the score for tapering has been developed. The index can be worked out using the formula 1=1.80+4.3 t, where 1 is the lesion size expressed in 1000 Cm^2 and it is the score for tapering. This will be useful in comparing the disease severity in different locations/cultivators as well as for a meaningful comparison of the efficacy of treatments (Mathew *et.al.*, 1989).

Control Measure : Calixin @ 25 g per 25 1 of water applied once in four months was effective in controlling the disease (Radhakrishnan, 1990).

4. Bud Rot

A brief account on budrot disease caused by *Phytophthora Palmivora* has been presented. Symptoms, causative organism, spread of the disease and its control have been included in the description (George and Rajeev, 1986). Bud rot disease often tends to be fatal. Manifestation of the diseases requires a combination of favourable factors of whch high relative humidity and low temperature are most impotant factors. Number of 'favourable days' determined the incidence and severity of the disease which in turn was dependent on the monsoon rains. Palms 3-20 yr. exposed to suitable micro climate were the most susceptible.

The role of *Phytophthora palmivora* (But) in the incidence of the diseases, the nature of damage, and the possible implication of secondary invaders, such as bacteria in aggravating symptoms have been assessed. Production of oospores in infected tissues suggests the mode of survival of the pathogen in the cabbage of affected palms. Under favourable conditions the infection cycle could be completed in about a week (Radha and Joseph, 1983).

Control Measure : The infected tissues from the crown region should be removed and protected with Bordeaux paste. Bordeaux mixture at 1% may be sprayed so as to reach the crown region as premonsoon spray.

Ridomil Mz 72 cumen L. mixture (101362-28-1) (0.5 + 0.08%) Bordeaux mixture (Boll-63-0) controlled bud rot effectively caused by *P. palmovora.* Fycap 40A (1332-40-7) and Fytolan (1332-40-7) were less effective than the other two fungicides. (Radhakrishnan, 1986).

5. Leaf Blight or Grey Blight

Leaf blight or grey leaf sopt a disease caused by the fungus *Pestalotia palmarum* cke affects the growth and vigour of coconut. The quality of seedlings is affected and in adult palms, nut yield is reduced.

The characteristic symptoms include appearane of minute yellow spots on the leaflets which later turn brown. Center of the spots turn greyish white with brown to dark margins. Many spots coalesce in to irregular grey necrotic patches. The fruiting body of the fungs is seen as black dot like structure. In advanced stage, tip and margin or the whole leaflets dry and shrivel giving the leaf a burnt appearance. This condition is generally referred to as 'leaf blight'. (Karthikeyan *et.al.*, 1994).

Fourteen fungicides were tested *in vitro* for their relative efficacy in controlling the grey blight of cocnut, caused by *Pestalotia palmarum*. Bavistin, hexathir and tecto-60 completely inhibited the growth of *P. palmarum* in all the concentration (100, 200, 300, 400 and 500 ppm). The growth of the fungus was also completely inhibited by brestan-60 (Fentin acetate) at 300, 400 and 500 ppm, captan at 400 and 500 ppm as well as difolatan 80 w and Dithane M 45 (Mancozeb) at 500 ppm. (Das and Mahata, 1985).

Twenty eight varieties and four hybrids were screened for grey blight disease of coconut. Among the hybrids, East Coast Tall x Dwarf Green had a fairly low disease index of 17.0 as againt 21.9 in East Coast Tall. Among the varieties, Chowghat Dwarf Orange, Dwarf Green, San Blas and Laccadive Micro showed relatively less incidence of grey blight. (Bhaskaran and Ramanathan, 1983).

6. Button Shedding and Premature Nut Fall

Heavy button shedding and premature nut fall cause enormous yield reduction in coconut. These are caused by several factors such as genetic variability, physiological or environmental conditions, fertilization, excessive soil moisture and drought, nutritional defficiency, diseases and pests. Various fungi *Phytophthora, Collectotrichum, Botryodiplodia, Botryosphaeria* and *Pestalotia* were reported to cause sheddingof buttons and immature nuts in coconut.

A fungus *Cochliobolus hawaiiensis* was found to cause button sheedding and premature nut fall in various coconut gentypes. The effects were pronounced in Lakshdweep micro (79.5%), Lakshadweep ordinary (54.6%) and East Coast Tall x Malaysian Green Dwarf (43.3%). Four month old buttons alone were found to be infected by the fungus of which, one month old buttons were highly susceptible to pathogenic infection. Fungicides mancozeb 0.2%, Carboxin 0.1% and Carbon oxychloride 0.3% concentrations recduced the severity of the disease.

7. Nutritional Disorders in Coconut

Deficineccies for nutrientts such as nitrogen, phosphorus, potassium and micronutrients are vey comon in coconuts cultivated in sandy loam soils. Symptoms are often observed in coconut gardens where the trees are not properly fertilized especially with organic manures. If highly deficient, the yield will be affected drastically and recovery will be possible only after two or three years. Hence a complete knowledge is necessary for preventing such disorders in coconuts.

Nitrogen deficiency manifests as uniform chlorosis (pale yellow) of lower leaves. The intesity of chlorosis is more at the older leaves which decreases progressively towards the younger leaves. Under acute deficiency, the affected older or lower leaves become pink or brown in colour and then droop down. The nuts are small in size with low kernel content.

Potash deficiency is manifested by scorching on the margins or the lower leaves. The trees exhibit wilting symptoms. The nuts become slender and narrow with very low kernel content.

The root growth is restricted if phosphorus deficiency is recorded. The kernel thickness will also be reduced with very low oil content. The vigour of the tree will also be reduced.

Marginal chlorsis of the older leaves is the characteristic defInciency symptom of magnesium. Initially, the margins of the leaflets turn pale yellow and later become yellow. The yellow colour progressively increases towards the midribs. However, the leaf lamina around the midrib remains green.

Zinc deficiency is characterised by formation of small leaves wherein the leaf size is reduced to 50%. Leaflets become chlorotic, narow and reduced in length. In acute deficiency, flowering is delayed. Zinc deficiency will also lead to button shedding.

Boron deficiency is exhibited by barren nuts. The nuts become long and narrow in shape. The kernel content will be reduced in thickness. Sometimes, the nuts will not contain any kernel due to acute deficiency. the terminal leaf will not expand and the leaflets become spindle and twisted.

Management practices include growing green manure (sunhemp) as an intercrop and ploughing *in situ*. Application of 650g urea, 1 Kg super phosphate and 1 Kg potash per tree along with 25 Kg farmyard manure (FYM) once in six months in June-July and December - January. Application of micro-nutrients viz., magnesium sulphate 150 g, zinc sulphate 75 g, borax 50 g along with 10 Kg of well decomposed FYM per tree prior to rainy season.

These measures help overcome nutritional disorders and restore the vigour and productivity of the trees (Balakrishnan *et.al.*, 1998).

8. Tapering Wilt or Pencil Point Disease

The affected palms will have the stem tapering like a pencil point towards the crown. Such trees contain very low number of leaves which often turn yellow with reduced size and less number of leaflets. Shedding of button and premature nuts is associated with the disease. The productivity is very much reduced and such palms die prematurely.

This disorder is caused by micronutrient deficiency and the trees could be restored to normal yield if, in addition to the application of major nutrients, the following micronutrients are applied every year at least for three years. Copper sulphate, zinc sulphate, manganese sulphate, ferrous sulphate and borax at 225 g each and ammonium molybdate at 10 g dissolved in 40 litres of water and all applied in the basin, 1.5 m radius around the tree followed by covering with top soil. The response will be good if applied in the early stage of the disease. Palms in the advancd stage of the disease may be cut and burnt. (Ramanathan, 1987).

References

Abdurahiman, V.C.; Mohamed, U.V.K. and Devi, O.K. (1982). Studies on the biology of a predator *Cardiastethus* sp (Hemiptera Anthocoridae) found in the galleries of *Nephantis serinopa* Meyr (Lepidoptera : Xylorictidae). *Curr. Sci.* **51** (11) : 574.

Abraham, V.A., Koya, K.M.; Abdulla and Kurian, Chandy, (1989). Integrated managemet of red palm weevil Rhynchophorus ferrugineus F.) in coconut gardens.. *J.plantn crops* **16** (suppl). : 159-162.

Abraham, V. A. and Mohandas, N. (1988(a). Biology of coconut white grub *Leucopholis coneophora* Burm (*Melolonthinae : Scrabaeidae : coleoptera*) *J. plantn. crops.* **16**(1) : 38-44.

Abraham, V.A. and Mohandas, N. (1988 (b). Chemical control of white grub *Leucopholis coneophora* Burm, a pest of coconut palm. *Trop. Agric.* **65**(4) : 355 - 357.

Abraham, V. A. andMohandas, N. (1990). Effect of different food plants on growth and development of coconut white grub *Leucopholis coneophora* Burm. *J. Plantn. Crops.* **18** (suppl) : 277-280.

Advani, Ranjan, (1984). Ecology, biology and control of black rat, *Rattus rattus* in Minicoy island *J. plantn. crops.* **12** : 11 - 16.

Anthony. J., Daniel Mariamma; Kurian, Chandy, and pillai, G.B. (1979). Attempts on introduction and colonization of the exotic reduviid predator, *Platymeris laevicollis* Distant for the biological suppression of the coconut rhinoceros beetle, *Oryctes rhinoceros,* L. In : Symposium on plantation crops, 2nd, Ootacamand, June 26-29, 1979. proceedings of PLACROSYM II Edited by C.S. Venkataraman and others pp. 445-454.

Balakrishnan, K., Nanthakumar, S. and Azakiamanavalan, R.S. (1998). Nutritional disorders in coconut. *'The Hindu'*, November, 5,1998.

Beena, C.G. 1980. Mating and ovipostion behaviour of *Trichospilus pupivora* Ferr (Hym : Eulophidae) *J. sci. Res.* **2**(2) : 115-117.

Bhaskaran, R., Ramadoss, N. and Ramachandran, T.K., 1987. Plant extract on the growth of *Ganoderma lucidum.* Tamil Naud Agric. Univ. News Letter **17**(7) : 2.

Bhaskaran, R., Ramadoss, N. and Ramachandran, T.K. (1988). Biologoical control of Thanjavur wilt disease of coconut. *Indian cocon. J.* **19** (6) : 3-8.

Bhaskaran, R. and Ramanathan, T. (1983). Variability of coconut varieties and hybrids to natural infection of grey blight disease. *Madras Agric. J.* **70**(10) : 685-686.

Bhat, S. Keshava and Sujatha, A. (1988). Effect of saturation baiting using multiple dose anticoagulants on the population of black rat in coconut fields. *Indian Coconut J.* **18** (9) : 1-4.

Bhat, S. Keshava and Sujatha, A. (1989(a)). Relative acceptance of different baits by *Rattus rattus* Wroughtoni Hinton. *J. plantn. crops* **17**(2) : 121-137.

Bhat, S. Keshava and Sujatha, A. (1989(b)). Laboratory evaluation of two anticoagulants against *Rattus rattus* Wroughtoni Hinton, *J. plantn. crops* **16** (suppl) : 217-221.

Bhat, S. Keshava and Sujatha, A. (1989(c). Prevent rat damage to tender coconut by bromadiolone baits. *Indian cocon. J.* **20** (7) : 3-5.

Bhat, S. Keshava and Sujatha, A. (1991). Toxicity and field efficacy of bromadiolone against *Rattus rattus* wroghtoni Hinton. *J. plantn cop* **18** (suppl) : 285-290.

Bhaskaran, R. Suriachandraselvan, M. and Ramachandran, T.K. (1990). Ganoderma wilt disease of coconut, a threat to coconut cultivation in India. *Planter* **66** (774) : 467-471.

Das, C. M. and Mahata, I.C. (1985). Evaluation of some fungicides agianst *Pestalotia palmarum* cke, incidence of grey blight of coconut. *Pesticides* **19**(5) : 37-39.

Deshpande, V.P.; Devaiah, M.C.; Govindan, R. ; Thipeswamy, C, and Goud, R. Rajeshekar, (1982). Observation on coconut shot hole borer, *Xyleborus semiopacus* Eichnoff (Coleoptera Scolytidae) *Curr. Res.* **10**(7) : 114-115.

George and Rajeeve, P. (1986). Bud rot of coconut palm; a disease caused by phycomycetes. *Indian Cocon. J.* **17** (4) : 13.

Ghorpade, K.D. 1977. A new species of *Pseudoscymnus* (Coleoptera Coccinellidae) predacious on coconut scale in peninsular India. *J.Nat. Hist.* **11** (4) : 465-469.

Ghoropade, K.D. (1979). On the association of some coccinellidae (Coleoptera) with spider nests. *Curr. Res.* **8** (6) : 105-106.

Ghosh, S. M. and Abdurahiman, V.C. 1984. Bioethology of *Meteridea* hutsoni (Nixon) (Hymenoptera : Braconidae), a parasite of *Opisina arenosella* walker, the black headed caterpillar pest of coconut. *Entomon* **9** (31) : 31-34.

Ghosh, S.M. and Abdurahman, V.C. (1985(a). Bioethology and evaluation of comparative efficiency of *Apanteles taragamae*, a larval parasitoid of the coconut caterpillar, *Opisina arenosella* - *Philipp. J. Cocon. studies* **10**(2) 4-8.

Ghosh, S. M. and Abdurahiman, V.C. (1985(b)). On some hyper parasitoids of *Opisina arenosella*, the black headed caterpillar pest of coconut in Malabar, India. *Cocos* **3**:29-31.

Ghosh, S.M. and Abdurahiman, V.C. (1987). A comparative study on the efficacy of entomophagus insects associated with the coconut caterpillar, *Opisina arenosella* walker. In : Advances in Biological Control Research in India. Edited by K.J. Joseph and V.C. Abdurahiman. pp. 92-101.

Ghosh, S.M. and Anbdurahiman, V.C. 1988. Some hyperparasitoides of *Opisina aranosella* Walker, the black headed caterpillar pest on coconut. *Philipp. J. Cocon. studies.* **13**(2) : 11-17.

Gopinathan, P.B. ; Mohandas, N. and Vasudevan Nair, K.P. (1990). Cytoplasmic polyhedrosis virus infecting red palm weevil of coconut. *Curr. Sci.* **59** (11) : 577-580.

Gowda, P.V. (1987). *In Vitro* studies on the interaction of antagonistic mycoflora with *Thielaviopsis paradoxa,* the causal agent of coconut stem bleeding disease. M.Phil dissertation. Mangalore University, Mangalore. 215 P.

Guruprasad, B.K. and Srihari, K. (1983). Rat damage and its control in coconut groves. In: Coconut research and developement; Proceedings of the international sysmposium 1st. Kasaragod. Dec. 27-31, 1976, edited by N.M. Nayar, New Delhi : Wiley Eastern, P. 393-397.

Jalaluddin, M. and Mohanasundaram, M. (1989 (a). Biological studies on *Aspidiotus destructor* Sign. a pest of coconut in Tamil Nadu. *Indian Cocon. J.* **19** (11) : 14-19.

Jalaluddin, M. and Mohanasundaram, M. (1989 (b)). Seasonal abundance of coconut scale *Aspidiotus destructor* sign. and its natural enemies in Tamil Nadu, India. *Indian Cocon J.* **20** (3) : 12 -17.

Jayaraman, V. (1985). Study on repellant action of phorate against rhinoceros beetle *Oryctes rhinoceros* Linn. In : Behavioural and physiological approaches in pest management. Edited by A. Raghupathy and S. Jayaran, TNAU, Coimbatore pp. 116-119.

Jeyaraj, R., Tahngaraj. T.; Vasuki, C.A. and Aruchami, M. (1986). Characteristics of phenol oxidases in larval cuticle of the coconut pest *Oryctes rhinoceros. Proc. Ind. Acad. Sci* (*Ani Sci.*) **95** (6) : 739-744.

John, T. Thomas, (1988). Control of mealy bugs in coconut gardens *Indian cocon. J.* **19**(5) 10.

Kannan S. and Prabhu, V.K.K. (1985). Probable role of brain-neurosecretary cells in water balance in *Oryctes rhinoceros* (Coleoptera : Scarabaeidae). *Entomon* **10**(4) : 313-315.

Kapadia, M.N. (1981). Susceptibility of three cultivated coconut varieties to *Nephantis serinopa* Meyr. *Indian Cocon. J.* **12** (4) 4-5.

Kapadia, M.N. (1982). Note of laboratory evaluation of some insecticides against the coconut leaf eating caterpillar *Nephantis serinopa* Meyr. *Indian Cocon. J.* 13(3) : 10-20.

Kapadia, M.N. (1983). Record of pediobius sp., hyper parasite of *Perisierola nephantidis,* M. and *Bracon brevicornis* W. in Gujarat. *Indian cocon. J.* **14**(5) : 7.

Kapadia, M.N. (1984). A new record of coconut scale insects and their enemies in Gujarat. *India cocon. J.* **14**(11) : 9-10.

Kapadia, M.N. (1987). Biology of cardiastethus sp hr. *nazarenus* Reuter (Hemiptera : Anthocoridae), a predator of *Opisina arenesella* Walk. *Gujarat Agric. Univ. Res. J.* **12**(2) : 55-56.

Karthikeyan, A. (1998). Early detection of coconut wilt. THE HINDU, July, 9, 1998.

Karthikeyan, A. ; Bhaskaran, R. and Suriyachandraselvan, (1994). Controlling leaf bilight and buttonshed in coconut. THE HINDU, October, 5, 1994.

Kashyap. R.K.; Verma, A.N. and Bhanet, J.P. (1984). Termites of plantation crops and their damage and control; review article. *J. Plantn. crops* **12** (1) : 1-10.

Koshy, P.K.; Sosamma, V.K. and Radhakrishanan Nair, C.P. (1975). Preliminary studies on *Radopholus similis,* infecting coconut and arecanut palms in South India. *Indian J. Nematol.* **5**(1) : 26-35.

Koshy, P.K.; Sundrararaju, P.; Sosamma, V.K. and Ravikumar K. (1985). Efficacy of four systemic nematicides agianst *Radopholus similis* in coconut nursery. *Indian J. Nematol.* **15**(2) : 148-151.

Kurian, Chandy; Abraham, V.A. and Koya, K.M. Abdulla. (1979). Studies on *paradasynus rostratus.* Dist. (Heteroptera : Coreidae), a pest of coconut. In : symposium on lantation crops, 2nd. Ootacamand, June, 26-29, 1979. Proc. PLACROSYM. II. Ed. C.S. Venkataraman *et.al.*, pp. 484-503.

Kurian, Chandy; Abraham, V.A. and Ponnamma, K.N. (1984). Attractants, an aid in red palm weevil management. In : Symp. Plantn. crops. 5th CPCRI, Kasaragod, Dec. 15-18, 1982. Proc. Placrosym, V Ed. K.V.A. Bhavappa *et.al.*, pp. 581-585.

Kurian, Chandy.; Natarajan, P. and Koya, Abdulla, K.M. (1977). Efficacy of two indigenous traps for trapping rodents in a coconut plantation *J. plantn. crops.* **5**(2); 71-74.

Louis, I. Henry and Chandrasekharan, N.R. (1980). A note on the weevil *Diocalandra stigmaticollis* as pest of coconut plam in Tamil Nadu. *Indian cocon. J.* **11**(11) : 5-6.

Mathew, Jacob,; Nambiar, K.K.N.; Jose C.T. and Anil Kumar, (1989). Stem bleeding disease of coconut; a method for indexing the disease severity. *J. Plantn. crops* **17**(2) : 80-84.

Mathew, K.P.; Nair, M.R.G.K. and Abraham, C.C. (1979). Toxicity of newer insecticides to coconut caterpillar *Nephantis serinopa* Meyr. and its parasite *Bracon brevicornis* wesmeal. *Agric. Res. J. Kerala* **16**(2) : 188-192.

Mohamed, U.V.K.; Abdurahiman, V.C. and Ramadevi, O.K. (1982). Notes on two carabid (Carabidae; coleoptera) predators of

Nephantis serinopa Meyrick (Xylorictidae : Lepidoptera) from Kerala. Entomon. **7**(3) : 341-343.

Mohan, K.S. ; Jayapal, S.P. and Pillai, G.B. (1983). Baculovirus disease in Oryctes rhinoceros in Kerala. *J. plantn crops* **11**(2) : 154-161.

Mohan, K.S.; Jayapal, S.P. and Pillai, G.B. (1989). Biological suppression of coconut rhinoceros beetle *Oryctes rhinoceros* L. in Minicoy; Lakshadweep by *Oryctes baculovirus* impact on pest population and damage. *J. plantn. crops.* **16**(suppl): 163-170.

Murthy, V.T. Sundara and Santhanakrishnan, K. 1978. Utility of unmated females of *Parasierola nephantidis* in the bilogical control of Nephantis serinopa Meyrick. *Curr Sci.* **47**(23) : 924-925.

Muthukrishnan, P. and Rangarajan, M. (1976). Laboratory studies on the control of black-headed caterpillar *Nephantis serinopa* Meyr. by micro-organisms. *Lab dev. J. Sci. Technol.* Part B. *Life Sci.* **12**(3) 106-108.

Nambiar, K.K.N. ; Joshi, Y.; Venugopal, M.N. and Chandramohan R. (1986). Stem bleeding disease of coconut; reproduction of symptoms by inoculation with *Thielaviopsis paradoxa. J. plantn. crops* **14**(2) : 130-133.

Pillai, G.B. and Bhat, S. Keshava, 1986. Parasitism development of progeny and sex ratio of *Parasierola nephantidis,* Mues. (Bethylidae) on different instars of *Opisina arenosella* wlk (*Nephantis serinopa)* Meyr. *Indian cocon. J.* **16**(11) : 3-5.

Pillai, G.B. and Bhat, S. Keshava, (1987). Biology and feeding potential of *Parena nigrolineata* (Choudoir) (coleoptera : carabiadae), a predator of the coconut caterpillar*Opisina arenosella* wilk. *Indian Cocon. J.* **17**(12) : 6-10.

Pillai, G.B. and Ramachndran Nair, K. (1981). Role of pupal parasitoides in the natural supression of the coconut caterpillar, *Nephantis serinopa.* Meyrick. *J. Plantn. crops.* **9**(2) : 84-87.

Pillai, G.B. Ramachandran Nair, K. (1982 (a)). A note on *Trichospilus pupivora* Ferr; a parasite of *Nephantis serinopa* Meyr. *Entomon.* **7**(2) : 225-226.

Pillai, G. B and Ramachadran Nair, K. (1982(b)). A technique for laboratory multiplictin of *Brachymeria nosatoi* Habu and other species of chalcidid parasitoids of *Nephantis serinopa* Meyrick. *Entomon* 7(2) : 213-217.

Pillai, G. B and Ramachandran Nair, K. (1990). *On the biology of Calleida splendidula* (Coleoptera : Carabidae) predator of the coconut leaf eating caterpillar *Opisina arenosella* Wlk. *Indian cocon. J.* **20**(12) : 14-17.

Ponnamman, K.N. (1984). Field studies on the comparative efficacy of BHC, dichlorvos and Trichlorphon for the control of *Opisina arenosella* wlk. *(Nephatis serinopa) Indian Cocon. J.* **15**(7) : 3-5.

Radha, K. and Joseph, Thomas. (1983). Bud rot disease of coconut. In : Coconut Research and Development; Proc. International symp. 1st. CPCRI, Kasargod, Dec. 27-31, 1976. (Ed) N.M. Nayar, New Delhi; Wiley Eastern. pp. 333-340.

Radhakrishnan, T.S. (1986). Control of bud rot caused by *Phytophthora palmivora* (But.) of coconut using newer fungicides. *Pesticides* **20**(1) : 45.

Radhakrishnan, R. (1987). Some field observations on the infestation of mealy bug and its management problems in coconut. *Indian cocon. J.* 17(11) : 14-25.

Radhakrishnan, T.C., (1990). Control of stem bleeding disease of coconut. *Indian cocon. J.* **20**(9) : 13-14.

Rahalkar, G.W. ; Harwalker, M.R.; Rananavare, H.D. ; Kurian, Chandy, Abraham, V.A., and Koya, K.M. Abdulla. (1977). Preliminary field studies on the control of the red palm weevil *Rhynchophorus ferrugineous* oliv., using radiosterilised males. *J. Nucl. Agric. Biol.* **63**(3) : 65-68.

Ramadoss, N.; Gunasekaran, M. and Ramanathan, T. (1986(a)). Chemical control of Thanjavur Wilt of coconut. In : Seminar on Management of Soil Borne Disease of crop plants (SOMSDOC), Jan. 8-10, 1986. Abstr. P. 21.

Ramados, N.; Gunasekaran, M. ; Vijayaraghavan, H. and Ramanathan, T. (1986 (b)). Studies on the destructive sampling of Thanjavur Wilt affected coconut. In : Seminar on Management of Soil Borne Diseases of crop plants (SOMSDOC). Jan, 8-10. 1986.Abrstr. P. 12.

Ramanathan, T. (1987). Technology for inceasing coconut production in Tamil Nadu. *Indian Cocon. J.* **17**(10) : 3-8.

Ranjith, A.M.; Razak, M.P. Abdul and Sumangala., S. Nambiar, (1988). Influence of weather on the larval population of cockchafer beetles in coconut. In : National Seminar on Agrometeorology in plantation crops. RARS, KAU, Pilicode, March, 12-13, 1987. Proc. pp. 83-89.

Rao, A., Ganeswara,; Rao, P. Ramamohan; Lakshminarayana K., and Rao, P. Sanjeeva, (1978). Prelimanary observations on the biology of *Parena laticincta* Bater (Coleoptera : Carabidae), a predator of *Nephantis serinopa* Meyrick. *Indian coconut J.* **9**(1) : 2-5.

Rao, A. Ganeswara, Rao, P. Ramamohan, and Rao, T. Ramamohan. (1989). Studies on the effect of root feeding of systemic insecticideds in the control of red palm weevil, *Rhynchophorus ferrugineus* Fab. in coconut. *Indian cocon. J.* **19**(9) : 12-16.

Rao, A. Ganeswara.; Rao, P. Ramamohan.; Thammi Raju, N.B.; Lakshminarayana, K.; Rao, B.H. Krishanamurthy and Murthy, Sathyanarayana. (1980). Administration of systemic insecticide through root – a new method of control of coconut black headed caterpillar, *Nephantis serinopa*, Meyr. *Indian Cocon. J.* **11**(1) : 3-5.

Rao, A.M.K. Mohana.; Subbiah, K.S.; Mekote, V.L. and Rao, A. Ramachand. (1983). Efficacy of crown baiting with warfarin and bromadiolone against rodents in coconut palms in Krishan District of Andhrapradesh. *Indian Cocon. J.* **15**(1) : 6-9.

Sadakathulla, S. and Ramachandran, T.K. (1990(a)). Evaluation of insecticides against termite, *Odontotermus obesus* Ramb (Termitidae : Isoptera) attacking palm and plaited leaves of coconut. *Indian cocon J.* **21**(1) : 11-13.

Sadakathulla, S. and Ramachandran, T.K. (1990(b)). A novel method to control rhinoceros beetle, *Oryctes rhinoceros* L. in coconut. *Indian cocon. J.* **21** (7-8) : 10-12.

Sankaralingam, A. and Usman, K. Mohamed, (1996). Root disease of coconut. THE HINDU, July 4, 1996.

Santhosh Babu, P.B. and Prabhu, V.K.K. (1987). How many larval instars are there in *Opisina arenosella? Entomon* **12**(1) : 39-42.

Sathiamma, B.; Jayapal, S.P. and Pillai, G.B. (1985). Record of *Ankylopteryx octopunctata candida* (Fabricius) (Neuroptera : Chrysopidae) as egg and larval predator on *Opisina arenosella* wlk; the leaf eating caterpillar of coconut palm. *Curr. Sci.* **54**(21) : 1128.

Sathiamma, B.; Pillai, G.B. and Jayapal, G.B. and Jayapal, S.P. (1986). Bioecology of *Bracon hebetor* say. (Hymenoptera : Braconidae) an indigenous larval parasitoid of the coconut leaf eating caterpillar *Opisina arenosella* wlk. (Lepidoptera : Cryophasidae). *J. plantn. crops.* **14**(2) : 140-146.

Solomon, J.J. and Govindankutty, M.P. (1991). Etiology Mycoplasma like organisms. In : Coconut root (wilt) disease. (Ed) M.K. Nair *et al.*, Kasaragod, CPCRI. Monograph series No. 3 pp. 31-40.

Sosamma, V.K.; Koshy, P.K. and Rao. E.V.V. Bhaskara, (1988). Response of coconut cultivars to the burrowing nematode, *Radopholus similis. Indian J. Nematol.* **18**(1) : 136-137.

Sreekumar, S. and Prabhu, V.K.K. (1988 (a)). Digestive enzyme secretion during metamorphosis in *oryctes rhinoceros* (Coleoptera : Scarabaeidae) *Proc. Ind. Acad. Sci.* (Anim. Sci.) **97**(1) : 67-71.

Sreekumar, S. and Prabhu, V.K.K. (1988 (b)). Probable endocrine role of midgut tissue in stimulation of digestive enzyme secretion in

Oryctes rhinoceros (Coleoptera : Scarabaeidae). *prcc. Ind. Acad. Sci.* (Anim.Sci.) **97**(1) : 73-78.

Sudheendrakumar, V.V.; Mohammad, U.V.K. Nerendran, T.C. and Abdurahiman, V.C. (1982). Biology and morphology of immature stages of *Bracon brevicornis* (Hymenoptera : Braconidae) an important bilogical control agent of black headed caterpillar, *N. serinopa,* a pest of coconut. *J. Bombay Nat. Hist. Soc.* **79** (2) : 254-260.

Sulladmath, V.V.; Shanthappa, P.B. and Ajanna, H.P. (1980). Etiology of stem bleeding disease of coconut in semidry tract of Karnataka. *Curr. Res.* **9**(5) : 73-75.

Sumangala, S. Nambiar, 1988. Susceptibility of hybrid coconut varieties to *Oryctes rhinoceros* Linn. (Scarabaeidae) under rainfed conditions at Pilicode. In : National symposium on coconut breeding and Management, KAU, Vellanikkara, Nov., 23-26, 1988.

Varkey, Thomas,; Kamalakshy Amma, P.G. ; Ramandan, P.L. and Nambiar, P.T.N. 1979. Foliar yellowing of cocnut palms in healthy and root wilt affected areas. *J. plantn crops* **7**(2) : 117-120.

Veeresh, G.K. 1983. Coconut grubs and their control in Karnataka, India. In : Pro. Coconut Res. and Dev. International symp. 1st. Kasaragod. Dec. 27-31. 1976. (Ed) N.M. Nayar, New Delhi : Wiley Eastern. pp. 385-391

Visalakshi, A., Beevi, S.N.; Pillai, K.S.; Nair, K.K.R. and Das N.M. (1987). Control of the Coreid bug pest on coconut. *Entomon.* **12**(4) : 395-396.

3

Oil Palm (*Elaeis Guineensis*)

INTRODUCTION

The cultivated oil palm began as a semi-wild crop in West Africa. Its oil entered international trade in the 19th century for the production of soap, margarine, candles and lubricants. The first plantations were started in the 1910s in response to increasing demand for its oil. The oil palm is now the second largest oil crop in the world, giving about 20 per cent (12 million tonnes) of the world's vegetable oil production entering international trade from about 3.3 million hectares of plantations (only 2% of the world's total oil crop hectarage). Malaysia and Indonesia account for significant area and production of oil palm. The other countries in which this crop is grown include Nigeria, Ivory Coast, Zaire, Cameroon, Papua, New Guinea and Latin America. (Soh, 1991).

Edible Oil Scene in India

Despite the impressive increase in oil seeds output from a level of 12-13 million tonnes till 1987-88 to 17 million tonnes at present, the demand for edible oil still falls short of the indigenous supply to the extent of a million tonnes and it has to be met through imports, thus causing a drain on the scarce foreign exchange of the country. Edible oils include groundnut, sesame, rapeseed-mustard, sunflower, safflower, soyabean and niger. Linseed and castor are non-edible oilseed crops.

Edible oil shortage could be reduced in the coming years by the introduction of oilpalm cultivation on a large scale and in the quickest time possible, oilpalm being the highest oil producer per unit area. Oilpalm is capable of yielding 4 to 6 tonnes of palm oil per hectare compared to less than a tonne fo oil from other cultivated crops (Iyer, 1991).

Origin and Spread

The botanical name of oilpalm is *Elaeis guineensis* Jacq. The palms are believed to have originated in Africa from where it spread to America and Far East. It was first introduced to Brazil in 15th century. The introduction of oil palm in South-East Asia began with four seedlings : two from Mauritius or Reunion and two from Amsterdam which were planted in Java in 1848 at the Botanical Garden, Bogor. In 1870, seedlings derived from these palms were planted in the Economic Garden, Singapore. In 1911 Deli Dura palms were brought and planted in Malaysia.

Introduction of Oilpalm in India

Oilpalm was introduced in India, as early as 1890 at National Botanical Garden, Calcutta. Subsequent introductions were made in several states in South, West and East India. The first effort of introducing oilpalm on a systematic basis was made by the Government of Kerala at Thodupuzha where a research station for oilpalm was set up in 1960. The growth of the palms at this station led to the large scale plantation development in Kerala in 1971. The palms and its products have received wide acceptance in South-East Asia. Rapidly increasing area under palms has helped to elevate the status of palm oil on the world vegetable oil market. There is ever increasing awareness for the cultivation of oilpalm in India (Abraham, 1991).

In recent years India has been exploring the possibility of cultivating oilpalm, as it is a rich source of edible oil. Though oil palm was introduced as a botanical collection, in 19th century at National Botanic Gardens, Calcutta, and later in Maharashtra during 1947-59, it was first raised as a plantation in 40 ha at Thodupuzha in 1960 by the State Dept. of Agriculture, Kerala. Oilpalm planting was taken up from 1971 to 1982 and an area of 3705 ha was covered by Oilpalm Corporation of India Ltd., Kerala. Yield obtained in 1971 plantation was 2.7 tonnes of oil/ha per annum. Experimental tenera palms gave an annual yield of 4.6 tonnes of oil/ha.

The Department of Biotechnology in collaboration with Governments in Andhra Pradesh, Karnataka and Maharashtra has taken up planting of 1000 ha each oilpalm during 1990-91. These plots in Krishna, East Godavari and West Godavari districts of Andhra Pradesh, Shimoga district of Karnataka and Sindhudurg district of Maharashtra were raised to demonstrate the feasibility of growing oilpalm under irrigated condition.

Other plantation include, 160 ha planted during 1987-88 in East Godavari district by 86 farmers, and 200 ha planted from 1988 onwards by M/s. Navabharat Enterprises in West Godavari district in Andhra Pradesh. A farmer in Quilon district of Kerala State has planted 81 ha in his tea estate. (Rethinam, 1992).

In Tamil Nadu, though the crop was introduced in Cauvery delta, in Tiruchi, Thanjavur and Thiruvarur districts, the uncertain water supply from the Cauvery, which is not a perennial river like those in Andhra Pradesh or Karnataka posed a problem. In the past four years (since 1994), out of a total of 1600 hectares under oilpalm, at present the cop survives only in 915 hectares. In 1997, the area brought under oil palm in the three districts was 750 hectares (Ganapathi, 1998).

Plant Products

The oil palm produces palmoil (83%) from the fuit pulp and palm kernel oil (17%) from the seed. Ninety per cent of the palm oil is used in food products such as cooking and frying oils, margarines, shortenings and confectionery fats. Non-food derivatives are mainly oleochemicals, e.g., fatty acids and dervatives, methylesters, fatty alcohols, obtained by chemically splitting the oil, which are used in the detergent, pharmaceutical, cosmetic, plastic, paper, textile, fuel and lubricant industries. Kernel oil is used mainly in oleochemicals, while the cake is used as animal feed. The fruit pulp contains 35% palmoil and the kernel 20% oil called kernel oil.

Palm Oil : Crude palmoil is deep orange/red and is semisolid at a temperature of 20°C. It is composed mainly of fatty acids which are present as glycerides. The main component fatty acids are palmitic acid (48%) and oleic acid (40%). It also contains 8% linoleic acid which is one of the essential fatty acids. Small quantities of beneficial non-fatty materials like carotinoids (650 ppm) are also present. The tocopherols (600 ppm), which have vitamin E activity, contained in the oil, protect it to some extent against oxidation that leads to rancidity. The sterol content is about 300 ppm.

Uses of palmoil : The main uses of palmoil are :

(i) In formulations for margarine and cooking fats such as Vanaspathi and shortening.

(ii) In the manufacture of biscuits, ice cream, soap, detergents and shampoos.

(iii) As a frying fat where its low foaming and longer keeping qualities stand out.

(iv) As a replacement in formulations of blends of hydrogenated liquid oils, marine oils and other oils and fats.

(v) Palmoil can also be used to yield vitamins, protein and antibiotics by fermentation process using selected micro-organisms.

(vi) Its inherent fatty acids are valuable byproducts which can be hydrogenated to give high quality stearic acid.

Palm kernel oil, the oil extracted from kernel of the fruits also has a variety of uses in margarine, shortening, cooking oils, biscuit fillings, biscuit fats, ice cream, cosmetics, soaps, shampoos, tube oil, additives, biodegradable detergents and a number of other products. But kernel crushing is costly and so the kernels are generally used as fuel in the boilers or may be converted to charcoal or otherwise used for metalling roads.

Taxonomy and Morphology

Oil Palm is a member of the monocot family palmae and belongs to the tribe cocoineae. There are three species under the genus *Elaeis*.

1. *Elaeis guineensis* (Commercial red oilpalm - African oilpalm).
2. *Elaeis oleifera* (American oilpalm).
3. *Elaeis odora* (Grown is Amazon).

Fig. 3.1 : Oilpalm tree

The economic life of oilpalm is about twenty five years with an immature period of about three years.

The principal morphological components of a mature oilpalm are : (1) root system, (2) shoot system, (3) inflorescene and (4) fruit. Abraham, V.K (1991).

Root System

The oil palm is a typical monocotyledon having a fibrous root system. Primary, secondary, tertiary and quarternary roots are the main components of this system. Under the primary root system, roots of 6-10 mm diameter emerge in thousands from the bole of the trunk. Individual primaries were found to a distance of 19 metres from the stem and absorbing portions of root were found at all intermediate distances. It spreads horizontally and vertically at varying angles into soil. The descending primaries are fewer in number and are intended for anchorage in the soil. All other old roots are strong and help in anchorage of the palm. Secondary roots of 2 to 4 mm, diameter are initiated from the primary ones. The secondary roots carry tertiaries of about 0.7 to 1.2 mm in diameter and in turn carry quarternaries. Quarternaries which are unlignified are often assumed to be the main absorbing roots and in adult palm it will be found within a radius of 3.5 to 4.5 meters. Apart from quarternaries, the tips of primaries, secondaries and tertiaries also absorb the nutrients from the soil.

Oilpalm root system except for the descending primaries, lies fairly close to the surface. The majority of the active root system is between 5 and 35 mm deep. Tertiary roots are in the upper 30 cm of soil. On deep soils the roots may be seen at still lower depths. Roots generally ramify extensively in soils without any concretionery materials. In soils of good physical properties, entire top soil is well permeated with roots. The effective absorbing roots are tertiary and quarternary ones in soil. In well drained loamy soils, the primary roots may reach upto 1.5 m downwards and 1.8 m horizontally within one year of planting in the field. The highest concentration of secondary roots is found in a zone 2 to 3 metres from the base of the tree.

The Shoot System

Trunk : The oilpalm has an unbranched habit producing a single trunk growing to a height of 15 to 18 metres bearing a crown of fronds (leaves). The girth of the stem (trunk) without leaf bases varies from 20 to 75 cm. The stem grows 35 to 75 cm in height each year. Internode length varies from 14 to 25 cm. The trunk remains covered by old leaf bases until the palm is about 11 to 15 years old. At this time the leaf bases start to fall usually from the middle of the trunk and extending downwards and upwards. The trunk of an old palm will usually be completely free of leaf bases except just below the crown.

Typically the stem is cylindrical in shape except for the bulbular bole at the base. The apical meristem lies in a slight depression at the apex of the stem. This meristem gives rise to all the vegetative parts of the crown. Stem tissue is formed at a later stage from a meristem which is continuous with the base of the fronds. The stem has three functions. First, it supports the leaves. Secondly, the vascular system embedded in the stem serves to transport mineral nutrients and water from the roots upwards and the products of photosynthesis from the leaves downwards.

Thirdly it also functions as a storage organ. The stem contains large amount of non-fibrous parenchymatous tissues. The estimated dry matter for trunks in a hectare (140 palms) will be about 84 tonnes. Oilpalm trunks take about 5 to 6 years to decompose. On account of this, the decaying trunk will harbour oilpalm pests like rhinocerous beetle and will become a source of various fungal diseases. Of late, trunks before felling are poisoned to increase the rate of decomposition and to avoid insect and disease incidence. In some estates in Malaysia the felled trunks are cut into small pieces with a chopper and then pulvarised using a specialised chopping machine. The pulvarised trunks are then distributed in the field as mulch and source of nutrients. Though oilpalm grows to a height 18 metres, at replanting stage, the trunk will have a height of 9 to 12 metres with a diameter of 45 to 60 cm. The trunk tapers gradually towards the crown. Oilpalm trunks and fronds are generally considered as waste. However, research is now directed to find out possible conversion of the trunk and frond to value added products such as pulp and paper, panel products, lumber and solid fuels.

Fronds (Leaves) : Some 45 to 50 leaves are found in the crown at a time. Each leaf remains enclosed for two years and then rapidly develop into a central spear and finally opens. The base of the leaf completely encircles the stem apex and in the adult leaf the base is persistant as a strong fibrous sheath. The mature leaf is pinnate. The petiole may be as long as 1.2 metres and the rachis about 8 metres. Two kinds of spines are found in the leaves. They are fibre spines and mid-rib spines. The leaf stalks bear lateral spines on both sides. Individual leaflets are linear in shape ranging from 250 to 300 in number. The number of leaves produced annually varies between 30 to 40 at 5 to 6 years of age. Thereafter, the production declines to a level of 20 to 25 per annum. Oilpalm frond can be obtained at harvesting time and during replanting. The average dry matter of a frond is 115 kg at replanting time, i.e. 16 tonnes per hectare (140 palms). The annual availability of pruned fronds will be about 11 tonnes per hectare.

Inflorescence : The oilpalm is a monoecious plant producing separate male and female inflorescence on the same palm in cycles. Normally not more than one inflorescence on a palm will be at anthesis at any one time. So self pollination is relatively rare. Each female inflorescence contains several thousand flowers and the receptive period for pollination is 36 to 48 hours. The flowers are produced in an inflorescence which is a compound spike. It develops from the axils of the leaves. The spikelets which are thick and fleshy are arranged spirally around the central axis. There will be about 85 to 165 spikelets in an inflorescence. The flowers are subtended by a bract. The overy is tricarpellary and androecium is rudimentary. Stigma is sessile, having three hairy lobes, exuding moisture. The male inflorescence has a long peduncle unlike female inflorescence. The spikelets or otherwise called as "fingers" have a length of 10 to 20 cm bearing 700 to 1200 flowers. The flower has a perianth with six segments. Androecium is tubular with six or seven anthers. Gynoecium is rudimentary. The flowers open and remain receptive from two to five days. Insects and wind are the natural pollinating agents. Harvested female inflorescence bearing the ripe fruits are known as fresh fruit bunches (FFB). Each palm bears about 10 to 12 bunches per year. The number of fruits per bunch varies from 50 to 100 in young palms and 1000 to 3000 in mature palms and the average weight of each bunch is 10 to 30 kg.

Insect Pollination in Oilpalm : The oilpalm hitherto thought to be wind pollinated, has been now proved to be an insect pollinated species. From West Africa, the original home of oilpalm, eight species of pollinating weevils were reported (Syed 1979). They belong to the order *Coleoptera* (Family – *Curculionidae*, sub-family – *Derelominae).* Weevils on oilpalm in cameroon include, *Elaeidobius Kemerunicus* Fst, *E. plagiaus* Fab, *E. singularis* Fst. *E. bilineatus* Fst. *E. subvittatus* Fst. and rarely *E. spatulifer* Mshl. The only species found in S. America is *E. subvittatus.* Occurrence of *E. Kamerunicus* in the oilpalm plantations of Kerala was introduced during 1985 and got established in Little Andamans in 1986. (Pillai and Ponnamma, 1992).

Fruits : Oilpam fruit is a sessile drupe, varying in shape from nearly spherical to ovoid or elongate. They are 3-5 cm long and weigh about 3.3 g. The pericarp of the fruit consists of the outer exocarp or skin, oil bearing mesocarp and the hard stony endocarp or shell. The endocarp together with the kernel inside forms the seed. Externally the fruit varies considerably at maturity. The exocarp of the external fruits of the bunch tends to be more pigmented than that of the internal fruits. The common type of fruit is deep violet or black at the apex and colourless at the base before ripening. Such fruits are called Nigriscens.

The fruit usually ripens five to six months after pollination. Depending upon the carotine content of the mesocarp there is variation in fruit colour. Three types of fruits are distinguished based on fruit colour.

Fig. 3.2 : Oilpalms tree with male and female inflorescences in the leaf axils

1. *Nigriscens* : Dark purple or almost black when young and reddish orange when ripe. (Most common commercial fruit type).
2. *Virescens* : Green when young, changing to bright reddish on ripening (Rare variety).

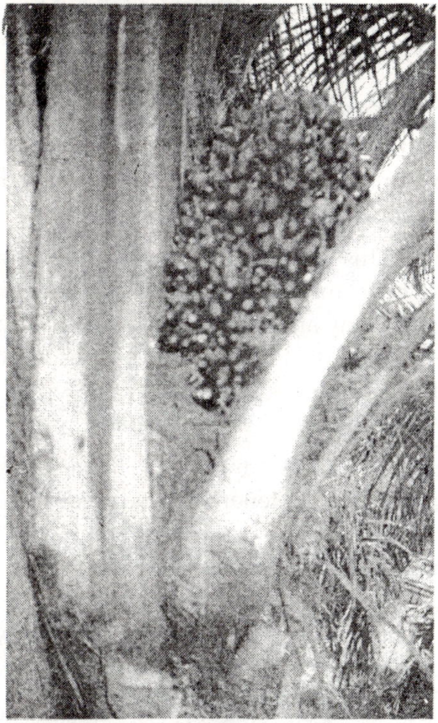

Fig. 3.3 : A fruit bunch in the leaf axil

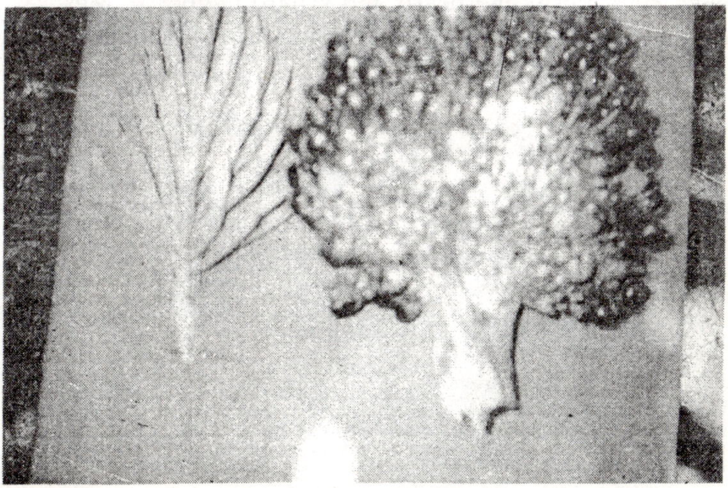

Fig. 3.4 : (*left*) Male inflorescence, (*right*) Female inflorescence

3. *Albescens :* Very dark brown when young and yellow when ripe (commercially not important).

There are three types of fruits based on thickness of shell :

(a) *Dura :* It is hard shelled. The average type has a shell of 2 to 8 mm, thickness; relatively little oil bearing mesocarp, thick endocarp and large kernel.

(b) *Pisifera :* This variety is almost shell-less mesocarp 92 to 97 per cent and kernel 3 to 8 per cent. The kernel is extremely tiny and the pulp very thick while the fruit is generally small. Pisifara has a high degree of female sterility and consequently bears only a few fruits. As such its commercial value is almost nil.

(c) *Tenera :* When the dura crosses with pisifera (D x P) a heterozygote, viz. tenera is produced. It has a thin shell (3 mm) surrounded by a ring of fibres in the mesocarp. In tenera variety this results in rich oil bearing mesocarp (60 to 90 per cent) compared to the thick shelled dura fruit. Tenera is the variety now commercially accepted as the best available planting material.

Fruit is sessile drupe. Fruits are either spherical or elongated, weighing about 50 g. The pericarp of the fruit consists of outer exocarp, a fleshy mesocarp and a hard endocarp. Mesocarp contains palmoil and the kernel inside endocarp contains palm kernel oil.

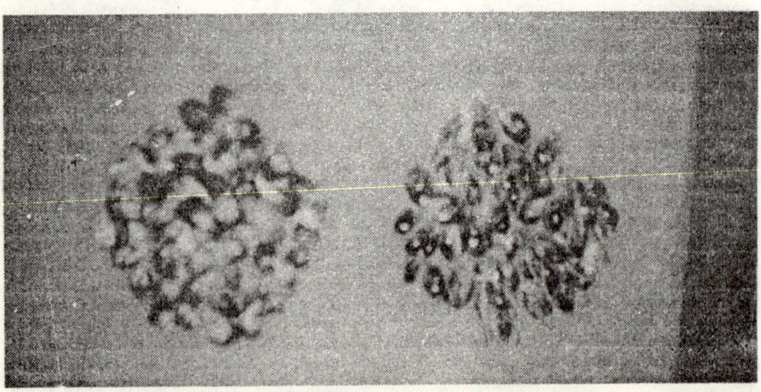

Fig. 3.5 : (*left*) Ripe fruits (Orange colour) (*right*) Unripe fruit (Dark blue colour)

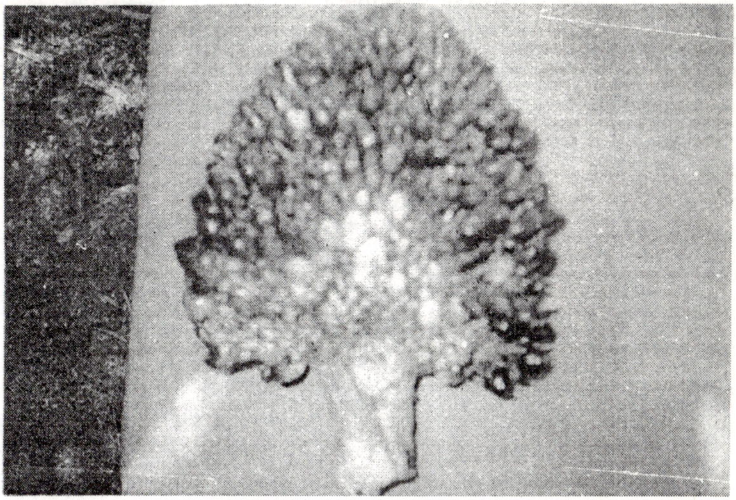

Fig. 3.6 : Ripe fruit bunch

Seed : The seed of oil palm is a nut. It consists of a shell, the endocarp and kernel inside. The shell has three germ pores, of which only one pore is functional. The embryo is embedded in the endosperm. The seed germinates in about hundred days after seed treatment and it ultimately becomes on oil palm seedling.

Seed Germination : The fruits are separated from the bunch and seeds are extracted by scaping off the exocarp and mesocarp with a knife, or by retting in water or by other mechanical means. In retting, the fruits are immersed in either running or still water for 10 days. The process is completed by pounding the fruits mixed in a wooden mortar to remove the adhering mesocarp. In dura fruit, a clean seed is obtained while in tenera a portion of fibres remain attached to the seeds. The seeds are then dried by spreading them on concrete or wooden floors under shade for two days. Such seeds can be stored for 3-9 months at about 27°C without much reduction in vaibility.

Under natural conditions, oilpalm seed takes about 2-2½ years to germinate. For quicker germination the following technique is adopted.

Seeds are soaked in water for five days changing the water daily. Thereafter, the seeds are spread out for drying for 24 hours.

The dried seeds are put in polythene bags (of conventional size and guage) closed tightly using rubber bands, and placed in a germinator maintaining a temperature of 40°C. After 80 days, the seeds are removed from polythene bags, soaked in water for 5 days changing the water daily and dried in the shade for two hours. After this, the seeds are put back into bags and kept in a cool place in order to maintain the moisture content. The germination commences in about 10-12 days. The percentage of germination obtainable by this method is 90-95%.

Raising Nursery : Of the different systems of nursery techniques, the single stage polybag system is the best. In this system, the polybag (preferably black) of 400-500 guage, measuring 40 x 35cm are used. The bags are filled with top soil or compost arranged at a spacing of 45 cm square and one seed per bag is dibbled. A good mulch during summer is desirable. Watering the seedlings is done thrice in a week. A fertilizer mixture containing 15g N, 15g P_2O_5 and 6g K_2O at the rate of 8g in five litres of water for 100 seedlings may be applied when the seedlings are two months and eight months old.

Planting : Oil palm is planted in the main field in triangular sysem at a spacing of 9m accomodating about 150 palms/ha. Planting is preferably done at the onset of the rains during June. The polythene bag is tore open and the entire ball of earth is buried in the pit (50 cm^3) and levelled. The young seedlings are protected from rodents wherever necessary, by a wire netting (45cm x 120 cm) encircling it at a radius of 15 cm from base.

Genetic Resources

The successful breeding of different oilpalm varieties to date has been founded on the availability of genetic variability for various traits of interest. The role of genetic resources will become even more important, in the future with new production and end use requirements and new genetic engineering technology. Extensive collection of wild *guineensis* and *oleifera* germplasm have been made by research institutes in various producer countries. (Raja naidu, 1986a, b; Meunier 1989, Escobar, 1981; Santos *et.al.*, 1986).

At CPCRI, Kerala, the germpalms holding consists of four exotic and one indigenous accession. The hybrid seedlings (teneara) produced by crossing dura palms of the oilpalm station, Thodupuzha with the pisifera pollen imported from Nigerian Institue for Oilpalm Research were planted over an area of about 15 hectares at Palode during 1976 for experimental purpose. It was observed that out of the different cross combinations planted 65d x 30.103 p hybrid gave the highest yield of 164 kg fresh fruit bunch (FFB) per palm per year followed by 120d x 30.103p with 159 kg. These two combinations have consistently given good yield over the years. The per hectare oil yield in the first case worked out to 4.6 tonnes (at 20% extraction rate) under rainfed conditions. The yield and fruit characters of some of the hybrid combinations are furunished in Table 3.1.

Table 3.1 : Yield and fruit characters of promising tenera hybrids (1986)

Cross combination	FFB/Palm/Yr.		Estimated Oil yield	Percentage to fruit		
	No.	Wt(Kg)	t/ha/yr.	Meso Carp	Kernel	Shell
65d x 30.103p	12.3	164.1	4.59	44.6	8.9	7.6
120d x 30.103p	12.0	159.2	4.46	47.7	6.6	10.9
271d x 30.4336p	12.8	146.4	4.10	56.8	4.8	8.6
61d x 30.4336p	11.1	138.5	3.88	50.1	6.0	10.7
125d x 30.103p	12.2	125.7	3.52	50.3	8.4	9.4
92d x 30.3154p	8.6	124.7	3.49	46.4	7.8	9.7

Till 1982, it was necessary to import the entire requirement of tenera seeds, since the male parent pisifera was not known to occur in India. A total of 25 pisifera palms have since been identified from the segregating populations of tenera hybrids. Promising dura (female parent) palms have also been isolated enabling CPCRI to produce 2-3 lakh tenera seeds annually (Bavappa, 1986(a)).

Breeding Objectives

Yield potential : Current oilpalm varieties have high yield potentials as evidenced by reports of record experimental and commercial fresh fruit bunches (FFB) yield exceeding 40 tonnes per hectare per year. Such record yields, however, are seldom sustained for more than one year. Whether this is a genotypic or agronomic constraint has yet to be ascertained. In any case, given the low heritability of the trait and the intensive selection made previously, significant quantum breeding improvements in FFB yield are unlikely unless perhaps palm architecture can be changed (Squires, 1984; Corley, 1986) which may require a long and perhaps uncertain breeding effort. The oil content of the bunch is more amenable to selection being more heritable. Current varieties average 24-25 per cent while exceeding 30 per cent are frequently found in breeding materials.

Besides oil content, oil palm breeders are now giving more attention to secondary traits such as oil quality and composition.

Oil Quality : Fatty Acid Composition

Palmoil has about 50 per cent saturated and 50 per cent unsaturated fatty acids (Table 3.2). Palmoil has made considerable inroads into the markets of other vegetable oils. Besides being cheaper, its stability conferred by low linoleic acid content and presence of tocopherols and tocotrienols makes it an ideal frying fat for making chips, instant noodles and snack foods. Its physical properties also allow its processing into margarines and shortenings without hydrogenation.

Table 3.2 : Fatty acid compisition of palmoil in breeding material of different origins (Noiret and Weridart, 1976)

Type of material	Fatty acid					
	Saturated			Unsaturated		
	Myristic C14:0	Palmitic C16:0	Stearic C18:0	Oleic C18:1	Linoleic C18:2	Iodine value
Deli	1.1	42.2	4.9	40.6	11.2	54.3
LaMe	0.4	36.4	8.6	43.4	11.4	56.8
Yangambi-Sibiti	0.9	45.8	5.2	34.4	13.7	53.3
NIFOR	0.7	42.1	6.8	39.7	10.7	52.7
Deli x LaMe	0.7	40.0	6.7	41.2	11.4	55.2
Deli x Yangambi	1.2	45.3	4.7	35.9	12.9	53.4
Deli x Sibiti	1.2	43.6	4.6	38.9	11.7	54.0
Deli x Nifor	1.1	43.1	5.0	39.6	11.2	54.8
(LaMe x Deli) x Deli	1.0	42.4	5.6	39.6	11.5	53.5

Fearing further competition, promoters for soyabean oil labelled palmoil as a saturated oil, consumption of which would lead to a rise in blood cholesterol levels predisposing to cardiovascular heart diseases. Promoters of palm oil counterclaimed that palmoil does not behave like a saturated oil because it is low in hypercholesterolemic fatty acid i.e. lauric and myristic, and very rich in the hypocholesterolemic monounsaturated oleic acid. Further more it has anticancer properties because of its rich B – Carotene (Vitamin A) and tocopherol and tocotrienol (Vitamin E) contents.

The subject of relative nutritional properties of different edibile oils is steeped in controversy, particularly in the health conscious affluent socieities. Some breeders prefer not to genetically change the fatty acid composition of palmoil as they believe that it is already a very versatile oil and chemical modification of the oil will be cheaper and faster. Others believe that the popular notion that polyunsaturated oils are better for health is entrenched in consumers' minds. Also there is a big advantage in producing a more liquid (unsaturated) palmoil to enable it to tap into the large cooking and salad oil markets in temperate countries.

Genetic variability for percentage unsaturation, expressed as iodine value is rather limited in advanced oilpalm breeding materials, although values of upto 66 percent can be obtained in slections of prospected semi-wild palms.

Drought Tolerance : Most commercial oilpalm is grown in favourable environments. Nevertheless, as oilpalm cultivation extends into less favourable or marginal areas, special adatable varieties will be needed. Most West African varieties have some level of drought tolerance as they have been bred in environments with a pronounced dry season. Such varieties or their derivatives would be suitable for other drought prone areas, e.g. Southern Thailand.

Breeding Methodologies

Population Improvement : Most breeding programmes have now adopted some form of population improvement, i.e. recurrent selection, which is a logical development as selected parents are interbred. Further, some inbreeding within a population invariably occurs, requiring interpopulation testing for efficient selection. This then becomes reciprocal recurrent selection (RRS).

The modified RRS scheme (Meunier and Gascon, 1972) has been adopted by many national breeding programmes which is logical as it allows a progressive improvement of the breeding populations yet maintains genetic variability. Improved lines can then be exploited for varietal production by public or private companies.

The recombinant inbred variety breeding scheme, based on the promise that hybrid vigour is not a consequence of heterozygosity and that by repeated selfing and selection, the superior F1 hybrid performance can be recaptured if not surpassed, has been proposed for the oilpalm (Pooni *et.al.*, 1989). The scheme is not likely to be adopted as the main breeding scheme of most programmes because of reservations about the number of generations needed and possible loss of genetic variability and yield associated genes as a consequence of severe inbreeding depression. It may be adopted in special programmes and perhaps for the development of new inbred lines.

Marker Assisted Selection : From association studies, Kouame and Noiret (1982) suggested that high mitochondrial activity could be used as a selection marker for increased yield. This technique has been adopted in the Ivory Coast. Other groups, however, are not so convinced.

RELP Technique : From the field of molecular biology, restriction fragment length polymorphism (RFLP) is now much touted to be a powerful selection tool. Although RFLP markers for many economic traits, even quantitative traits, have been found in many annual crops, their use in routine breeding programme is still somewhat limited because of the high cost and time needed to screen the large numbers of genotypes turned over per generation. In oilpalm breeding where large genotypes numbers and rapid turn over of generation are not so critical, a case may be made for RFLP marker assisted selection for some more heritable economic traits such as short stature, oil quality and disease resistence in specific programmes.

Transgenic Palms : The requisite to realise transgenic oilpalms, i.e., knowledge of the biochemistry of the desired trait, isolation and cloning of the desired gene, transfer and incorporation of the gene into the plant cell, plant regeneration and stable expression of the trait in the transgenic plant, will take years to meet, by which time the desirable traits currently envisaged for genetic engineering may have been achieved to some extent by conventional breeding. Undeniably, once the technuques are established the process will be much faster with subsequent efforts.

Also because of position effects, the engineered gene may not be stably expressed or have undesirable pleiotropic efffects which would require further breeding. Breeders will not be surpirsed if transgenic palms should mainly turn out to be novel means of increasing genetic variability, especially from unrelated species, for further breeding.

Clonal Multiplication : Crop improvement in oilpalm through conventional methods has been slow due to its perennial nature, long juvenile phase and high level of heterozygosity due to cross pollination and lack of conventional methods for vegetative propagation. The crop is essentially seed propagated and the commercial plantations are with hybrid tenera (dura x pisifera) palms. The use of hybrids comes in the way of establishing uniform plantations as the inbred parental lines are not available (Nwankwo and Krikorian, 1982). Recently much emphasis has been given to the oil quality aspects as well. It is thus attractive to develop methods for the vegetative propagation inorder to establish uniform plantations of elite genotypes. There are many reports on the vegetative propogation of oilpalm through tissue culture (Jones, 1984; Rebechault and Martin, 1976, Paranjothi, 1984; 1986). Raju *et.al.*, (1989) have reported the development of clonal oilpalms from tender leaf explants of three year-old tenera seedlings through somatic embryogenesis with a minimum callus phase.

Ever since the success in clonal propagation of oilpalm was demonstrated by Jones (1974) much attention has been centered on the commercial exploitation of this method. Tissue culture method currently followed comprises the elaborate callus phase followed by shooting from meristemoides and rooting (Paranjothi, 1986). In contrast to the meristemoid formation which essentially needs separate shooting and rooting treatments for developing the plantlets (Wooi, 1984), the method described by Raju *et.al.*, (1986) produced embryoids from the compact calli at the lower cut end of the explant, very similar to the sexual embryos and germinate with both shoot and root. The germination of the embryoids was found to be 75 to 80%. The time taken for the production of the plantlet of 5 cm height in tube from the leaf explant was only four months. This time, however, was found to vary with the plant used for the initiation of the culture.

Somatic Embryogenesis : Oilpalm plantlets were produced through somatic embryogenesis after a very brief callus induction phase from tender leaf segments of *tenera* seedlings. The 5 cm long leaf bits placed on to the semisolid nutrient media and incubated in the light showed cellular proliferation at the cut margins in 2-12 weeks. The callus with compactly arranged densely cytoplasmic cells responded to somatic embryogenesis. In a period of 8-20 weeks depending on the source plant, somatic embryogenesis was observed. Somatic embroys appeared as small white club shaped structures and attained a size of 3-5 mm in length in 6-8 weeks with well defined shoot and root poles. Embryoids were independent structures and could be separated from each other even at 0.2 mm size as there were no vascular connections. They were found to multiply by a process of budding in the embryoid induction medium. These embryoids germinated with almost simultaneous shoot and root developement. The clonal plants after attaining a size of 5 cm were transferred to a potting mixture of sand and vermiculite moistened with Hoagland's solution fortified with IBA. Over 80% of the plants survived the transfer to potting mixture. (Bavappa, 1986).

A method of the non-destructive extraction of tender leaf tissues from elite mature trees was reportedly standardized. The leaves from one third circle of the crown were removed to get acess into the central column. The leaves from spindle onwards were cut leaving two to three whorls of outermost leaves intact. After clearing the leaf base upto the stem region, the central tender leaf column with the covering leaf sheaths, were exised at about 15-20 cm above the exposed stem. The leaves with in the column were surface sterilised and used as explants. The stump in the apex was treated with fungicides and insecticides. All the seven high yielding trees sampled for clonal propagation were reported to have survived and recorded good growth.

Embroyo Culture : The embryos of oil palm nuts of various maturity stages were cultured *in vitro* in MS and SH media containing auxins, NAA, 2, 4-D and cytokinins, BAP Kinetin and 2 iP to develop plants. The mature embryos started development soon after inoculation while the embryos from stored dry nuts showed a dormancy period of upto eight weeks in culture. In higher concentrations of 2, 4-D with cytokinis, callus formation was noticed at the base of the shoot around the root pole. The haustorium remained while throughout the culture period. The withdrawal of cytokinis completely from the medium resulted in rooting to get oilpalm plantlets (Bhavappa, 1986).

Earlier Breeding Efforts

Systematic breeding of the oilplam began in Africa in the period between the two world wars and has intensified since the 1950s. Until the late 1950s the thick-shell fruited *dura* variety Deli formed the base of this expansion, particularly in the Far East. This relatively uniform big fruit-bunched, high oil bearing 'Variety' was the result of early breeding efforts in Indonesia and Malaysia with four palms from Bogor, Indonesia, which presumably came from the same parent tree in W. Africa *Via* Amsterdam and Mauritius (or Reunion). (Hartley, 1988). There were also active breeding efforts on the promising grove palms in Congo, Ivory coast, Cameroon and Nigeria, giving rise to various superior breeding populations of restricted origins or BPROs (Rosenquist, 1986), e.g. Yangambi, LaMe and Calabar. When it was firmly established in the 1940s that the thin shelled *tenera* is a monofactorial hybrid of the thick-shelled *dura* and the shell-less female sterile *pisifera.* The switch to *ternera* or *dura* x *pisifera* cross as the commercial 'variety' proceeded accordingly. It was soon also established that the Deli *dura* when crossed to *pisiferas* of W. African BPROs gave heterotic effects, firmly establishing the Deli *dura* x W. African *pisifera* hybrids as the predominant commercial 'varieties'.

A breeding improvement of 65 per cent increase in oilyield over five generations (about 50 years) in Deli *dura* has been estimated. With the switch over to *tenera,* a further 20 percent increase was achieved. Breeding improvements in *tenera* was about 15-19 percent per generation (Hardon, *et.al.*, 1987).

Breeding has undoubtedly played an important role in the development of the industry in response to its initial needs which were essentially uniform, high yielding varieties. The oilplam industry is now a mature industry. With new discoveries and uses found for oil palm products, consumer requirments have become more sophisticated and discriminating while at the production end, declining land and labour availability and rising costs are undermining the competitiveness of the industry in some countries such as Malaysia, (Soh, 1991).

Special Quality Oils : In recent years, breeders have been asked to breed for higher contents of kernel oil for the oleochemical industry, of stearic acid for cocoa butter substitute production and of B- carotene, and tocopherols and tocotrienols for the production of vitamins A and E, respectively. Even if no premiums are guaranteed for such speciality oils, it will be prudent for breeders to include such traits in their breeding programmes in case future markets may demands oils with such attributes.

Breeding for Short Stature : Most current high yielding varieties grow at the rate of 45-65 cm per year so by 20 to 25 years the bunches are very difficult to harvest. Much of the yield decline in older palms can be attributed to inefficient harvesting of tall palms. Harvesting tall palms is a tough, skilful and largely manual operation with little prospect of cost effective mechanisation. Malaysia and other countries face an acute shortage of good harvesters. When short mutants or forms were first observed in some BPROs (breeding populations of restricted origin), their potential in facilitating harvesting and prolonging due to early preoccupations with high yields, abundunt cheap labour and purported deficiencies in the short forms, interest in them waned somewhat until recently.

The best known short palm is Dumpy. Besides imparting its shortness, it also imparts undesirable traits such as few large bunches, resulting in lower yield, and heaviest leaf bases, hindering pruning, to the progenies of its out-crosses. This led to loss of interest in it as a *dura* parent in many programmes. An organisation in Indonesia continued production of Dumpy *dura* based tenera varieties. Apparently, their growers are willing to trade a 20 per cent reduction in early yields for a 40 per cent reduction in palm height. Another organisation in Malaysia persisted in the incorporation of Dumpy genes into the *pisifera* population by back crosses to superior AVROS, BPROs and found that many of the undesirable Dumpy traits disappeared or were greatly ameliorated, yet shortness was retained (Soh *et.al.*, 1981). The commercial variety produced by corssing these dumpy - AVROS *pisifera* with Deli duras is about 20 per cent shorter yet comparable in yield to the superior Deli *dura* x AVROS *pisifera* variety (Soh, *et.al.*, 1981).

There is now more breeding interest in other short palms such as Gunung Melayu of Deli, 16R of Yangambi, Pobe 'Dumpy' LaMe, population 12 of PORIM's (Palmoil Research Institute of Malaysia), Nigerian genetic resource material and *E. Oleifera* which is extremely short.

Ideotype Breeding : Ideotype style breeding for increased physiological efficiency is emphasised in some programmes (Rosenquist, 1990). Physiological traits advocated for selection include leaf-area ratio (leaf area per unit weight) and leaf area expansion rates, relating to efficiency of light energy captured, econversion efficiency (efficiency of conversion of energy captured to dry matter) and harvest index (efficiency of conversion of dry matter to the economic product (Breure, 1985; Breure and Corley, 1983). While the theoretical concepts are appealing, most breeders are not likely to place great amphasis on these traits individually other than selecting for an overall intuitive or conceptual ideotype, which is essentially a high oil yielding palm with a short trunk and a full but light-weight conopy, in their main breeding programmes. Fristly, ideotype breeding has limited success even in annual crops (Rasmusson, 1987). Secondly, isogenic lines are not available in oilpalm to test the value of these traits, while supporting evidence based on correlation studies has been prone to self correlations (statistical dependencies) as most of the physiological parameters were derived variables in which yield or total dry matter formed a major component (Chang and Rao, 1989).

Breeding for Disease Resistance

1. *Stem Rot* : The oilpalm in the Far East is generally free of devastating diseases and pest. The only exception is basal stem rot caused by *Ganoderma boninense.* This was considered to be a serious problem mainly of older oilpalms planted on previous coconut or oilpalm land in areas with high water tables. Reports of the disease affecting younger palms on ex-oilpalm areas in inland soils are becoming frequent. Chemical and cultural controls have not been very effective. Until very recently, breeding for resistant palms has been hampered by lack of an efficient artificial inoculation and nursery screening technique as field screening is eneffective because the disease occurs sporadically.

2. *Vascular Wilt* : Breeding for tolerance to vascular wilt caused by the soil-borne pathogen *Fusarium oxysporium* s. sp, *elaedis* forms an important part of programmes in W. Africa, as the disease causes wide-scale loss of young and old oilpalm stands. Tolerant materials and a reasonably effective nursery screening technique are available. Apparently, derivatives of Dumpy possess high tolerance to this disease. The disease does not exist in Far East, although *F. oxysporium* occurs commonly. Despite strict phytosanitary restriction against importation of plant and soil materials from W. Africa, Asian programmes should include tolerant materials from West Africa besides Dumpy, as additional insurance.

3. *Sudden Wither and Lethal Bud Rot* : Serious diseases such as sudden wither or Merchitez caused by the *Phytomonas protozoan* and lethal bud-rot of uncertain causal origin have greatly impeded the expansion of oilpalm cultivation in South America. Resistant varieties appear to be the best form of control and resistance seems to be

found mainly in *olifera* x *guineensis* hybrids. South American breeding programmes may have to be committed to the development of *Oleifera* x *guineensis* hybrids to support their oilpalm development plans (Meunier, 1989).

References

Abraham, V.K. (1991). Oilpalm from root to fruit. *Indian Oilpalm Journal* **1** (3) : 15-20.

Bavappa, K.V.A., (1986(a)). Research at CPCRI, PP. 26-27.

Bavappa, K.V.A., (1986(b)). Biotechnology. In Research at CPCRI, pp. 56-58.

Breure; C.J. (1985). Relevant factors associated with crown expansion in oil palm (*Elaeis guineensis* Jacq) *Euphytica* **34** : 165-75.

Breure, C.J.; Corley, R.H.V. (1983). Selection of oilpalm for high density palnting. *Euphytica,* **32**: 177-86.

Chang, K.C.; Rao, V. (1989). Some comments on the use of the solar radiation conversion efficiency in oilpalm breeding, *SABRAO Journal* **21** : 151-54.

Corley, R.H.V. (1986). Yield potentials of plantation crops. In : Potassium in the Agricultural systems of the Humid Tropics IPI. Bern, Switzerland. PP. 61-80.

Escobar, C.R. (1981). Preliminary results of the collecting and evaluation of the American Oilpalm (*E. oleifera* HBK, Cortex) in Coasta Rica. In : The oilpalm in Agriculture in the eighties. Kuala Lumpur, Malaysia, pp. 79-93.

Ganapathi, V. (1998). Oilpalm cultivation catches on in delta, 'THE HINDU', October 6, 1998.

Hardon, J.J.; Corley, R.H.V.; Lee, C.H. (1987). Breeding and selecting the oil palm. In : Improving vegetatively propagated crops. Long Ashton Research Station, U.K. pp. 64-79.

Hartley, C.W.S. (1988). The oilpalm. Narlow, U.K.: Longman, London. pp. 761.

Iyer, N.S. (1991). Edible oil scene in India with special reference to palm oil. *Indian oilpalm journal* **1**(3) : 1-8.

Jones, L.H. (1974). Propagation of clonal oil palms by tissue colture. *Oil palms News* **17** : 1-8.

Kouame, B.; Noiret, J.M. (1982). A precious test for productivity in the oilpalm (*Elaeis guineensis* Jacq.) by measurement of mitochondrial activities. In : The oil palm in Agriculture in the eighties. Kuala Lumpur, Malaysia, pp. 33-46.

Meunier, J. (1989). Prospects of *E. olifera* x *E. guineensis,* interspecific hybrids. IRHO results. In : Proceedings of ISOPB/PORIM workshop on prospects of interspecific hybrids, PORIM, Malaysia pp. 23-27.

Nampoothiri, K.U.K.; Pillai, R.S.N.; Premkumar, T. (1984). Oilpalm pamphlet No. 19, pp. 1-18, CPCRI, Kasaragod, Kerala.

Paranjothi, K. (1984). Oilpalm. In : Hand book of Plant Cell Culture (Eds) P.V. Ammirato; D.A. Evans; W.R. Sharp and Y. Yamanda, Macmillan Press, New York, pp. 591-605.

Paranjothi, K. (1986). Recent developments in Cell and tissue culture of oilpalm. PROIM OCC, Papaer palmoil Res. Inst., Malaysia. No. 19, pp. 12.

Pillai, R.S.N. and Ponnamma, K.N. (1992). Botany. In : Oilpalm production technology. (Eds) Nair, M.K. and Nampoothri, K.U.K. CPCRI, Kasaragod, pp. 13-20.

Pooni, H.S.; Cornish, M.A.; Kerasey, M.J.; Lawrence, M.J. (1989). The production of superior lines and second cycle hybrids by inbreeding and selection, *Elaeis.* **1** : 17-30.

Rajanaidu, N. (1986(a). The oilpalm (*Elaeis guineensis*) Collection in Africa. In : Workshop proceedings. Palm oil Research Institute of Malaysia No. 10. pp. 59-83.

Rajanaidu, N. (1986(b). *Elaeis oleifera* collection in central and South America. In : Workshop proceedings. Palm oil Research Institute of Malaysia, No. 10. pp. 84-94.

Raju, C.R.; Sajini, K.K. Balachandran, S.M.; Saj K.V.; Geetha Maheshan, K., Rajasekharan, P.E.; Geetha, L.; Bavappa, K.V.A. 1986 Clonal multiplication of oilpalm (*Elaeis guineensis* Jacq), *J. palntn crops* **16** (Suppl) : 17-20.

Rasmusson, D.C. (1987). An evaluation of ideotype breeding. *Crop Sci* **27** : 1140-46.

Rebechault, H. Marthin, J.P. (1976). Multiplication vegetative du plamier a haile *Elaeis guineensis* (Jacq) a paide de cultures de tissue foliares. *C.R. Acad Sci. Paris 283* : 1735-37.

Rethinam, P. (1992). History of oilpalm cultivation and potential in India. In : Oil palm production technology (Eds) Nair, M.K. and Nampoothiri, K.U.K. CPCRI, Kasaragod, pp. 7-11.

Rosenquist, E.A. (1986). The genetic base of oilpalm breeding populations. In : Workshop proceedings. Palm oil Research Institute of Malaysia. No. 10, pp. 27-56.

Rosenquist, E.A. (1990). Improvement of tenera population using germplasm from breeding progremmes in Cameroon and Zaire.

In : Proceedings of workshop on progress of oilpalm breeding populations. pp. 37-69.

Santos, N. DEM.; Barcelos, E., Nascimento, J.C. (1986). Genetic resources of *Elaeis oleifera* H.B.K., cortes in the Brazilian Amazon. In : Workshop proceedings, Palmoil Research Institute of Malaysia No. 10 pp. 95-101.

Soh, A.C. (1991). Oilpalm Breeding - breeding into 21st Century. *The planter* **67** (782) ; 203-215.

Soh, A.C., Vannialingam, T., Taniputra, Panim, K. (1981). Derivatives of the dumpy palm some experimental results. The Planter **57** : 227-239.

Squires, G. (1984). Principles for selecting oilpalm ideotypes. PORIM Internal Report.

Syed, R.A. (1979). Studies on oilpalm pollination by insects. *Bull ent. Res.* **69** : 213-224.

Wooi, K.C. (1984). Palm tissue culture. In : Micropropagation by selected root crops, palms, citrus and ornamental spices. FAO plant production paper No. 59, FAO, Rome. pp. 88-112.

4

Protection Technologies of Oilpalm

PESTS AND DISEASES OF OILPALM

A. Pests

Oilpalm is infested by a wide range of pests including insects, birds and mammals. Some species of insect pests affecting coconut and aracanut also cause considerable damage to oilpalm. In adult oilpalms, the rhinoceros beetle, *Oryctes rhinoceros* L., and red palm weevil *Rhynchophororus ferrugineus* (Oliver) are the major pests, while Limacodids and Psychids causing frequent defoliation and coccoids infesting fruit bunches are minor pests. Termintes and pseudococcids and bag worm are other pests.

Rhinoceros Beetle : The rhinoceros beetle, *Oryctes rhinoceros,* a serious pest of coconut, was attacking oilpalm in recent years. The adult beetle which bores through into the spear leaves, resulting in snapping of the fronds. However, unlike in coconut, damage to inflorescence was not noticed in oilpalm. The effective period for treating the crown in oilpalm against rhinoceros beetle damage was April to June, as the leaf damage due to rhinoceros beetle was high during the above period. Crown application of napthalein balls (4 Nos.) covered in perforated polythene sachet or phorate granules (2g) in polythene sechet or neemseed powder and sand (1 : 1) mixture equally and effectively checked the leaf damage in oilpalm upto 3 months, (Anon, 1997). In nature, the rhinoceos beetle is suppressed by entomopathogens like *Baculovirus oryctes* (Virus) and *Metarhizium anisopliae* (fungus). Release of *Baculovirus oryctes* in the oilpalm plantation and *Metarhizium anisopliae* on the grubs of rhinoceros bettle in the breeding sites of the pest is recommended.

Red Palm Weevil : The damage caused by red palm weevil, *Rhyncophorus ferrugineous* (Oliver) is due to the feeding activity of the grubs, which bore through and feed on softer tissues of stem and meristem. Palms infested by red palm weevil show gradual wilting and drying of outer whorl of fronds and also rotting of spear.

Intensity of infestation of rhinoceros beetle and red palm weevil in oil palm is relatively less than that in coconut.

Other Insects : Unripe and ripe fruit bunches were infested by Coccoids (Pseudococcids, Diaspids and Coccids). They encrust the ripe and unripe fruit bunches. The pseudococcid, *Palmicultor* sp infested the spear cluster and unfolding leaves, resulting in yellowing of unfolded leaves and stunted growth of the palm.

Case worms (Psychids) and nettle caterpillar (*Limacodids*) are also infesting oil palm (Dhileepan, 1991(b). Defoliation caused by psychids and Limacodids was restricted mostly to outer whorl of fronds and occasionally in the middle whorl of fronds.

Insect Pollination in Oilpalm : Pollinating weevil *Elaeidobius kamerunicus* was used to increase pollination in oil palm. The activity of pollinating weevil and male inflorescence production at Aduturai (TN) was high during January to April and hence the release of pollinating weevil in oil palm garden (12 weevils/ spikelet) can be taken up during this period.

Release of pollinating weevil on oil palm garden increased the fresh fruit bunch weight by 26.2% and fruits to bunch percent by 15.3.

Bagworm : Oilpalm is widely grown in Nagapattinam, Thanjavur and Trichy districts of Tamil Nadu. Regular pest monitoring observations in the oilpalm gardens revealed bagworm, *Metisa plana* is a common pest. The other species which attack oilpalm is *Cremastopsyche pendula.* Bagworm caterpillars are characterised by their possession of a case which they inhabit throughtout their development. The case is made of silk to whcih, the caterpillars attach fragments of leaf tissues from the plant, on which they feed.

The capterpillars cause severe damage – the young larvae scarp the epidermis while the mature ones make holes in the leaf.

There is progressive necrosis of the leaf tissue and eventual skeletonisation. In young palms where the fronds are not in contact with other palms, the dispersion of the pest is usually restricted, which checks the outbreak of the pest and severity of the damage.

The larvae of *Metisa plana* feed for about eight weeks. The caterpillars are white with characteristic brown markings on the head and thorax. The head and thorax regions protrude from the case during feeding and locomotion, but when the caterpillar is resting or moulting they are withdrawn and the case is fastened to the plant surface. The case is composed of small pieces of leaf tissue which are left loose at the lower end of the case giving a ragged appearance.

The bags or cocoon cases are 15 mm long and are fixed to the leaf by a short crook shaped attachment. The males (with wings) emerge three weeks after pupation and fertilise the females which (wingless) always remain in her bag. The female lays 100-300 eggs and they hatchout in 2 to 3 weeks. The total life span is three months.

A number of parasites and predators attack these species. The most common parasite of the larvae is the wasp, *Apanteles* sp. The cocoons are attacked by many hymenopteran parasites, the most common species is *Brachymeria* sp. The predatory adult assessin bug is also reported to feed on the caterpillar.

The outer whirl of fronds harbouring large number of bagworms may be cut and burnt. New oilpalm gardens upto three years may be sprayed with any of the contact insecticides such as endosulfan whereas in gardens with an activity or pollinating insect (after three years), spraying of chemical insecticides should be avoided. As an alternative, preliminary studies on the management of the pest revealed that spraying of plant products such as neem oil 2% emulsion or notchi leaf extract 10% was effective in checking the damage and inhibiting further spread (Sridharan, *et.al.*, 1998).

Birds : Avian pests such as crows (*Corvus splendens protegalus, Corvus macrorhynchus culminatus*, 'mynah' (*Acridotheres tristis tristis*), and parrots (*Psitticula Krameri manillensis*) have been found damaging oilpalm fruits. Birds feed on the mesocarp and cause-fruitloss upto 2.8 t/ha/year (Dhileepan, 1989). Covering the bunches with different materials such as wire nets, reed baskets, plaited coconut leaf baskets and senile oilpalm leaf are effective in preventing the fruit damage. But senile oilpalm leaf covering is more practical and economical.

Rodents : Black rat (*Rattus rattus* wroughtoni), house rat (*Rattus rattus* rufescens), lesser bandicoot (*Bandicota bangalensis*) and larger bandicoot (*Bandicota indica*) are the rodent pests affecting oilpalm. Oilpalm seedlings and young palms upto the age of three years are more prone to rodents. Rodents tunnel into bole of young palms and kill them. Different baits such as iron livetraps, snaptraps, deathfall trap, bow trap etc. may be used to minimise rodent damage. (Dhileepan and Jacob, 1992).

B. Diseases

Oilpalm is affected by a number of diseases. Among these, only spear rot cause considerable economic losses. Early leaf diseases, leaf rot, leaf spots and nursery spear rot are the diseases affecting seedlings in the nursery stage. In adult palms, apart from spear rot, bud rot, crown disease, upper stem rot, leaf rot, leaf spots and bunch failure are the other diseases affecting oilpalm. Other devastating diseases like vascular wilt, fatal yellowing, suden wither and Ganoderma basal stem rot reported from other oilpalm growing countries have not been reported so far from India (Babu and Nair, 1992).

Diseases Affecting Nuresery

Stray incidence of leaf rot has been recorded in the primary nursery seedlings during rainy period. *Corticium solani* is the causal agent of this disease. The infection begins at the base of the spear as irregular pale olive green patches surrounded by brown zone, which later turn dark brown before drying. Removal and destruction of disease affected leaves and application of carbendazim 0.1% are recommended for controling the disease.

In the case of early leaf disease caused by *Botryodiplodia theobromae*, small light brown spots develop on the distal portions of the leaves, which later change to dark brown. These lesions enlarge and are surrounded by yellowish halo. The entire leaf tips get affected. Black dots (Pycnidia of the fungus) are visible on the lesions. Clipping of severly affected leaf portions and spraying with dithiocarbamates 0.2 effectively control the disease.

In nursery spear rot-reported in the seondary nursery, seedlings at Palode – water soaked lesions develop in middle leaflets of spear leaves. They later become light brown and extend to the tip of the spear. When the spear unfurls, the rotten area becomes dessicated. The young leaves show distinct marginal chlorsis comparable to the spear rot in adult palms. Subsequently emerging leaves show drastic reduction in leaf size. Such seedlings need to be culled out and destroyed.

Leaf spots caused by *Pestalotiopsis palmarum* and *P. monochaetoides* occur in older leaves of secondary nursery seedlings. Leaf spots caused by *Curvularia lunata* and *c. geniculatus* were recorded on primary and secondary nursery.

Adult Palm Diseases

1. *Spear rot* : The etiology of spear rot in India is yet to be established. Among the most consistent fungal isolates are *Fusarium moniliforme* and *Colletotrichum gloeosporioides. Bacillus subtilis* (Ehrenberg) was also found to be associated with the disease. None of these fungi and bacteria were pathogenic eigther individually or in combinations. Though the presence of Mycoplasma Like Organisms (MLOs) in the sieve tubes of rachillae of unopened inflorescence was indicated, the involvement of MLOs in the etiology of the disease is yet to be confirmed.

Yellowing of the youngest whorl of unfolded leaves is the initial visible symptom. Yellowing starts from the tips of leaves and spreads mostly along the margins of leaflets. The chloratic area later turns brown and dries up. In few cases, one or two affected leaves break at the leaf base and hang down from the crown. Yellowing and rotting does not spread to the older leaves in the whorl. Rotting and reduction in leaf size occur in the subsequently emerging leaves. With the desease advancement, rudimentary leaves even without leaflets are produced. As the disease advances, the trunk gradually tapers and inflorescences abort before emergence resulting total loss in productivity.

Macro and micronutrient status in diseased and healthy palms show no significant variation.

In view of the uncertain etiology, roguing of affected palms and burning *in situ* is recommended to prevent the spread of this disease.

2. *Bud Rot* : Rotting initiates at the basal portion of the spear closer to the meristem and extends to the whole spear. The spear could be easily pulled off. Higher disease incidence is noticed in young plantations. Cleaning the affected tissues and drenching the crown with carbendazim 0.1% cures the disease. The leaves emerging immediately after the application of fungicides are shorter and successively emerging ones are normal.

3. *Crown Disease* : During the early phase of the disease brown lesions with water soaked margin are observed on the middle portions of the internal leaflets. No rotting is apparent in the spear externally. When the affected spear unfurls, it develops a characteristic bending in the middle of the rachis. The leaflets on either side of the rachis are either completely rotten or a few stumps remain. The recovery of the disease affected palm is spontaneous and hence control measures are not taken up usually.

4. *Upper Stem Rot* : The symptoms are gum exudation, bleeding and rotting of the stem. Gum exudation and stem bleeding occur at any point above 50 cm from the bole region. Decay of inner tissue is observed on chipping of bark as the disease advances. The presence of healthy tissues above and below the lesions indicate that the infection is likely to be localised. Identification of the disease in the early stage and removal of rotten tissue, swabbing with calixin 0.1% followed by application of hot coal tar is recommended to control the disease.

5. *Leaf Rot* : Brown spots with yellow halos appear on the leaflets of inner whorl leaf lamina. Palms of all ages are found affected, but the severity is on palms up to ten years of age. Though *Colletotrichum gloeosporioides* has been reported as the casual agent, earlier inoculation trials could not establish the pathogenicity.

6. *Leaf Spots* : This disease caused by *Pestalotiopsis* spp are noticed on the lower whorls of leaves. Magnesium deficiency

favours disease incidence. The first symptom is the appearance of brown spots on leaflets, later spreading to the whole leaflets. The centre of the lesions are light grey to brown with numerous black dots, the acervuli of the fungus.

7. *Bunch Failure* : This is most frequent in 3-10 years old palms, often to the extent of 20 per cent. The tips of underdeveloped fruits turn black and later become necrotic. Retention of rotten immature bunches serves as a source of infection on fresh bunches. *Marasmius palmivourus,* a saprophyte usually colonises leaf axils and leaf bases. The bunches, in the middle stages of development are mostly invaded by the pathogen. Bunch failure may also result from lack of adequate pollination, in which case assisted pollination is to be resorted. Complete control could be obtained through crown cleaning, assisted pollination and bavistin 0.2% spray when applied together. Introduction of pollinating weevil, *Elaeidobius kamerunicus* reduces bunch failure incidence.

Processing Technology

Results of work done at Kumulur (Tamil Nadu) are presented here. (Anon, 1997).

Palmoil is an important vegetable oil in the diet of people in many parts of the world. It ranks second in world production of edible oil. As our country is beset with acute shortage of edible oil, oilpalm holds great promise.

Palmoil is extracted from oilpalm fruits through a series of unit operations namely sterilisation, stripping, digestion, extraction and clarification. Among the above unit operations, sterilisation and extraction are important operations. Sterilisation of oilpalm fruit is done to inactivate the lipase and lipolytic anzymes which are responsible for increase in FFA. Extraction of palmoil can be done by hydraulic press, screw press and solvent extraction methods.

The oilpalm fruits were sterilised at steam pressure of 1 kg/cm^2 for 75 min., 2 kg/cm^2 for 60 min., 3 kg./cm^2, for 45 min. and micro-wave sterilisation for 5 minutes. The effect of different sterilisation conditions on extraction efficiency of hydraulic press and FFA value of crude palmoil were studied.

A hand operated hydraulic press of 60 tonnes load capacity was used for extraction of palmoil.

The oilpalm fruits used in the present study had a mean size fo 3.72 x 2.64 x 2.48 cm and the sphericity and roundness were found as 0.78 and 0.54 respectively.

The principal factors affecting the extraction efficiecny of hydraulic press are, applied pressure, pressing temperature, holding time and different conditions of sterilisation. An attempt has been made to optimise, these parameters for getting the maximum extraction efficiency.

The optimum ranges of the variable for getting the maximum extraction efficiency were determined by conducting experiments designed according to RSM. The level of the factors taken for the study were,

Factor	Level
(i) Applied pressure P(Kg/CM^2)	70(-1), 75(0), 90(+1)
(ii) Pressing temperature, T(ºC)	80(-1), 90(0), 100(+1)
(iii) Holding time t (sec)	90(-1), 120(0), 150(+1)

The maximum extraction efficiency was obtained at microwave sterilisation (71.3%) at minimum holding time of 90 sec. followed by 3 ksc sterilisation (70.8%).

The mean volume, specific gravity and hardness of the raw oilpalm fruit were 10.44 cc, 1.06 and 50 g respectively. It was found that the volume and specific gravity increased during sterilisation while the hardness decreased. The maximum increase in volume and specific gravity and the maximum decrease in hardness was found in 3 Kg/Cm^2 sterilisation, while the reverse was observed in the case of microwave sterilisation.

The effects of the various factors viz. applied pressure (P) pressing temperature (T), holding time (t) on the extraction efficiency of hydraulic press and for 1 Kg/Cm^2, 75 min., 2 Kg/Cm^2, 60 min., 3 Kg/Cm^2, 45 min., and microwave sterilised oilpalm fruit can be described by the following first order multiple regression model.

E_1 = 9.043 + 0.252P + 0.229T + 0.55t

E_2 = 39.584 + 0.117P + 0.834T + 0.030t

E_3 = 47.782 + 0.145P + 0.079T + 0.023t

EM = 42.904 + 0.238P + 0.093T – 0.026t

Where E_1 = Extraction efficiency of hydraulic press for sterilised fruits at 1 Kg/Cm^2 pressure for 75 min.

E_2 = Extraction efficiency of hydraulic press for sterilised fruits at 2 Kg/Cm^2 pressure for 60 min.

E_3 = Extraction efficiency of hydraulic press for sterlised fruits at 3 Kg/Cm^2 for 45 min.

EM = Extraction efficiency of hydraulic press for sterilised fruits with microwave for 5 min.

P, T, t are pressure, temperature, time at which pressing of oil was done.

Maximum extraction efficiency was obtained from microwave sterilisation (71.3%) followed by sterilisation at 3 Kg/Cm^2 (70.8) at minimum holding time of 90 seconds.

In all the cases, the maximum extraction efficiency obtained was between 85 to 90 Kg/Cm^2 of applied pressure and 95 and 100°C of pressing temperature.

The effect of different packaging materials on FFA and peroixide value of crude palmoil was stuided for 60 days. White glass bottles, brown glass bottles, tin container and plastic bottles were selected for this study. The FFA and peroxide values were analysed for a constant duration of 15 days. Plastic bottles gave the poorest protection and tin container gave the best protection from oixdation. No significant effect of packaging material on FFA value was noticed.

The FFA value of oil (2.2%) under microwave sterilisation was significantly lower than the other conditions. The high FFA sontent of 8.2 percent was observed under 1 Kg/Cm^2, 75 min. sterilisation.

There was no significant effect of packaging material on FFA value of crude palmoil. However, oil stored in plastic containers showed high FFA value than the other material at the end of the storage period.

There was significant effect of packaging material on peridicle value of the crude palmoil. Oil packaged in plastic bottle developed a significantly higher peroxide value (23.6 M.eq/Kg) than the other material, whereas, oil in the tin container developed a significantly lower peroxide value than the other materials (10.7 M.eq/Kg).

References

Anonymous, (1997). Proc. Sixth Palm Scientists meet, 11th November, 1997, Aduturai, pp. 126-144.

Babu, M. Kochu and Nair, R. Radhakrishnan, (1992). Diseases and disorders. In : Oilpalm production technology pp. 59-69. (Eds) M.K. Nair and K.U.K. Nampoothiri, CPCRI, Kasaragod, Kerala, India.

Dhileepan, K. (1989). Investigation on avian pests of oilpalm. *Elaeis guineensis* Jacq. in India *Tropical Pest Management*, **35** : 273-277.

Dhileepan, K. (1991(b)). Insects assoicated with oilpalm in India *FAO Plant prot Bull.* **39**(2/3) : 94-99.

Dhileepan, K. and Jacob, S.A. (1992). Pests. In : Oilpalm Production technology pp. 49-58. (Eds) M.K. Nair and K.U.K. Nampoothiri, CPCRI, Kasaragod, Kerala, India.

Shridharan, S.; Jeyarani, S. and Subramanian, M. (1998). Bagworm damage in oilpalm. 'THE HINDU', June, 30, 1998.

Author Index

Subject Index